A HIKER'S GUIDE TO TRAILSIDE PLANTS IN HAWAI'I

John B. Hall

Photography by John Hoover and Ken Suzuki

with

Robert Aldinger, Thomas H. Rau, Vincent T. Soeda, Bradley F. Waters

MUTUAL PUBLISHING

All rights reserved
Photographs by Robert Aldinger, John Hoover, Thomas H. Rau,
Vincent T. Soeda, Ken Suzuki, Bradley F. Waters where indicated.
Cover photos, from top left, clockwise: John Hoover, Ken Suzuki, Karen Liliker,
Ken Suzuki, Thomas H. Rau
All other photos were taken by the author.

Design by Kyle Higa
ISBN-10: 1-56647-872-3
ISBN-13: 978-1-56647-872-4

Library of Congress Cataloging-in-Publication Data

Hall, John B.
  A hiker's guide to trailside plants in Hawaii / John B. Hall ; photography by John Hoover and Ken Suzuki with Robert Aldinger ... [et al.].
    p. cm.
  Includes index.
  ISBN 1-56647-872-3 (softcover : alk. paper)
  1. Coastal plants--Hawaii--Handbooks, manuals, etc. 2. Forest plants--Hawaii--Handbooks, manuals, etc. 3. Mountain plants--Hawaii--Handbooks, manuals, etc. I. Title.
  QK473.H4H25 2008
  581.9969--dc22

                    2008019594

First Printing, July 2008

Mutual Publishing, LLC
1215 Center Street, Suite 210
Honolulu, Hawai'i 96816
Ph: (808) 732-1709
Fax: (808) 734-4094
email: info@mutualpublishing.com
www.mutualpublishing.com

Printed in Korea

TABLE OF CONTENTS ■■■■■

Thomas H. Rau

DRY FOREST PLANTS

JOHN HOOVER

MESIC FOREST PLANTS

ACKNOWLEDGMENTS ■■■■■

I owe a great debt to all the people who have offered their companionship and shared my love of our mountain plants with me as we hiked, scrambled, and chopped our way through the rocks, mud, and brush of our rugged and beautiful hills. I especially include the members of the Hawaiian Trail and Mountain Club, Solemates, the Nature Conservancy, the Over-the-Hill Gang, the Sierra Club, the Audubon Society, HOOTS, Mabel Kekina's trail maintenance crew, and the unaffiliated hikers who have joined me on rambles through our wild lands.

Stuart Ball, whose detailed trail guides have encouraged so many novice hikers to sample our splendid back-country, must be mentioned in particular. I hope that this volume will find space beside his in many a day-pack. John Hoover initiated this book project, and has provided constant encouragement and guidance in manuscript writing, coping with the computer, and negotiating the business side of the publication process. He has also supplied a large fraction of the photographs needed to illustrate the manuscript. For many years Ken Suzuki has self-published elegant little booklets of plant photos that he has generously given to his friends and to the people on the tours he guides for the Hawai'i Nature Center to help them identify the plants that they see. He deserves a major share of credit in the present work for the many lovely photographs he has contributed to it. Many other friends have also donated much-appreciated pictures, and their names appear beside each photo they supplied. I particularly appreciate the patience and expertise of my son-in-law, Richard Uyeda, who, by a blessed fate, is a computer programer, for his ability and willingness to bring my computer to heel when it became whimsical or downright vindictive, and for guiding me through the mysteries of word processing and the manipulation of digital photographs. My debt to John Obata and Dan Palmer, as mentioned in the Introduction, is immeasurable.

Many other friends have contributed greatly to my enjoyment and knowledge of our plant life, including Gerald Carr, Loren Gill, Clyde Imada, Jim and Yuko Johnson, Bruce Koebele, Charles Lamoureux, Tom Mendes, Sue Monden, Ken Nagata, Kost Pankiwskyj, Tom Rau, Dan Sailer, Norman Scofield, Celia Smith, Roger Sorrell and David Frost, Brandon Stone, Mashuri Waite, Brad Waters, and Charlotte Yamane. Needless to say, sole credit for any errors, absurdities, or infelicities that remain in this work must be laid at my door.

John B. Hall

INTRODUCTION ■■■■■

When I moved to Honolulu in 1962, I immediately began hiking with the Hawaiian Trail and Mountain Club. Within a few years, I had hiked most of the trails used by the Club, taking an interest in the plants as a way of extending and enlivening the hiking experience. Pete Holt, a professional botanist, began my botanical education. Later, two expert amateur botanists, John Obata and Dan Palmer, taught me most of what I know about Hawaiian plants. After I retired, I trained as a docent for the Hawai'i Nature Conservancy and led groups through the Honouliuli Reserve. This work motivated me to learn more about the way that plants were involved in human affairs, ancient and modern, and about our impact on their survival and success.

This book is for hikers wanting to know more about the plants they encountered along Hawai'i's trails. A description of each plant, along with photographs supplied by my friends and collaborators will make it easy to identify them. There are also interesting facts—place in the ecosystem, legends, uses, and related plants. (More detailed information can be found in the *Manual of the Flowering Plants of Hawai'i,* revised edition, 1999, by W. L. Wagner, D. R. Herbst, and S. H. Sohmer.)

The plants include common natives as well as introduced weeds, and most are easily seen. Though fungi are not technically "plants," but are more closely related to animals, a few of the more noticeable ones, along with a small sample of mosses, liverworts, and an alga (which are plants) are included.

Whenever possible, I use both the common (or Hawaiian) and scientific names. Usually plants that have no ornamental or practical value will have no common or Hawaiian name. For them, I use the scientific name. A number of native plants are found in genera (groups of species) with several, or even many, species. Most Hawaiian names apply to all species in a genus, and in general, I will only identify a plant at this level, unless different species in the genus have features that are worth noting. Groups of related genera are called families. Family names, which indicate where the plant fits into the plant kingdom, always end in "-aceae," pronounced "ay (as in hay) –see–ee."

A word about Hawaiian names. Hawaiian vowels sound like vowels in Spanish. The consonants are quite straightforward, except for "w" which is sometimes pronounced as a "v," and the 'okina, or glottal stop, represented by '. It is pronounced like the hyphen in "Oh-oh!" The 'okina is a real consonant, and omitting it may change the meaning of the word. Hawaiians did not form the plural of a name by adding an "s" to the end. I follow the Hawaiian practice of using the name both singularly and plurally: one koa, two koa, etc. (Note that "species" is both singular and plural also.) The macron, a bar over a vowel, indicates that that vowel is somewhat longer and should be stressed.

The book is divided into climate zones: Coastal, Dry Forest, Mesic (intermediate) Forest, Wet Forest, and Alpine. Within each zone, plants are divided into Herbs, Shrubs, Trees, Vines, Ferns, and Miscellaneous. The Coastal zone is the most clearly marked, consisting of everything near enough to the coast to be affected by salt spray and the generally arid conditions. Dry Forest begins at the trailhead away from the coast and includes the area dominated by shrubs with a few alien trees such as ironwoods and silk oak. Guavas, Christmas berry, and lama occur here as stunted shrubs rather than full-sized trees. The Mesic zone begins when these trees begin to reach twenty feet or more in height. It extends to where the trail begins to be consistently muddy in the low, shady spots, even in dry weather; where 'uki (Machaerina angustifolia) a large sedge with long, glossy, strap-like leaves begins to appear on ridges; and where ha'iwale, (Cyrtandra spp.) becomes common in the gulches. Alpine zones are found only on the higher mountains of the Big Island or on Haleakalā on Maui. They are generally shrublands with a few small trees, inhabited by plants that tolerate low temperatures and bitter winds. Many plants grow in several zones, and are described in the first zone they appear. What I consider a tree, Christmas berry, for example, may often be a shrub in the driest area in which it is first met.

To make it easier to locate a plant, each climate zone is distinguished by a colored bar on the outer edge of the page: blue for Coastal, brown for Dry, red for Mesic, green for Wet, and gray-green for Alpine.

Readers are often interested in how rare or widespread a plant is. An "endemic" plant is one found here and nowhere else. It may be endemic to the archipelago, to an island, or to a particular spot on one of the islands. An "indigenous" plant arrived here without human assistance, but is also found elsewhere. A "Polynesian introduction" was brought to Hawai'i by the original Polynesian settlers. An "introduced," "alien," or "exotic" plant was brought in after the arrival of Europeans or Americans. Hawai'i has a larger percentage of endemic plants than any area of comparable size in the world. Unfortunately, many are extinct or in serious danger of becoming so. While there are efforts to reverse this trend, it is hard to be optimistic, given the constant influx of invasive weedy plants and destructive insect pests; the devastation wrought by feral pigs, goats, rats, slugs, insects, and disease organisms; as well as ever-spreading development and climate change.

A few more terms must be defined. A "simple" leaf is one in which a single leaf blade is attached to a twig by a stem. Some simple leaves are lobed, like a papaya or maple leaf. If the gap between lobes extends all the way to the midrib, and each lobe has its own stemlet, we have several discrete "blades" attached to the same stem. This is a "compound" leaf. The ʻūlei, for example, has a leaf stem to which pairs of small, oval leaflets are attached, on opposite sides, with a final single one at the end. This is called a "pinnately compound leaf"—"pinna," meaning "feather"—since the leaflets are arranged along the stem like the fibrils on the shaft of a feather. The coconut palm frond is a good example of this. Since there are an odd number of leaflets, the ʻūlei leaf is said to be "odd-pinnately compound." Some leaves are "palmately" compound. The stemlets to which the leaflets are attached converge to a point, like fingers on a palm, and then join the main stem all together. A fan palm or an octopus (or rubber) tree, *Schefflera,* have examples of this kind of leaf. "Opposite" leaves are attached opposite each other in pairs along the stem. "Alternate" leaves are staggered along the stem, with a leaf on one side lying between a pair on the other.

■■■■■

JOHN HOOVER

JOHN HOOVER

THOMAS H. RAU

Coastal Plants

HERBS

ʻAKULIKULI
Sesuvium portulacastrum
Aizoaceae (New Zealand spinach, lithops or living stones)

This indigenous plant occurs along coasts throughout the tropics. It is a succulent, sprawling herb with red or green stems that often root at the nodes. The paddle-shaped, opposite leaves are ¼ to 1 ½ inches long and narrow with rounded tips. They may turn yellow or red with age. Small attractive star-shaped 5-petalled flowers are usually pink, but may be white. The plant is also called "sea purslane". No uses by the early Hawaiians have been recorded as far as I am aware.

Look for: A sprawling, succulent coastal herb with small paddle-shaped leaves and pink 5-petalled star-shaped flowers.

ALENA
Boerhavia spp.
Nyctaginaceae (four o'clocks, bougainvillea)

There are 4 species of Boerhavia in Hawaiʻi, 1 introduced weed, 2 indigenous, and 1 endemic. They are all sprawling herbs of dry coastal environments and adjacent lowlands with opposite leaves well spaced along the stem, ½ to 1 ½ inches long, narrow to nearly round, on short stalks. The flowers are tiny and inconspicuous, generally pinkish in color, tubular, and 5-lobed. The slender flower stalks of the native species are generally

Look for: A low growing coastal herb with opposite leaves and tiny sticky oval fruits on long, slender, much-branched stalks.

3

short, 1 to 3 inches long, but those of the weedy introduced alena, the one you are most likely to encounter, are much longer, often a foot or more in length, and widely branching. The fruit is a small oval, about ⅛ inch long, and sticky, readily clinging to skin or clothing. The introduced species comes from the Caribbean and was first noticed in Hawai'i in 1974. The indigenous species are also found on islands in the South Pacific, one extending as far as Africa.

AUSTRALIAN SALTBUSH
Atriplex semibaccata
Chenopodiaceae (beet, spinach, quinoa, pigweed)

This plant is a sprawling, mat-forming herb with alternate leaves that are about ½ to 1 inch long. Leaves tend to have a fine white fuzz underneath and are pointed oblongs with a few irregular teeth along the edges. The stems are often reddish in color. The plant usually bears small pink fleshy fruits with 1 or more triangular teeth at the tips. This herb is native to Australia and was reportedly introduced to Lāna'i about 1895 as a potential cattle forage plant. It has since spread and is now found at low elevations along dry coasts of all the main Islands. It takes

Look for: A low-growing herb of dry lowland coasts with alternate, toothed leaves, small pink fleshy fruits, and a salty taste.

up salt from the soil in which it grows, and leaves and fruit have a salty flavor as a consequence.

HINAHINA
Heliotropium anomalum
Boraginaceae (kou, comfrey, heliotropes, forget-me-nots)

There are 4 species in this genus in Hawai'i, 2 introduced and 2 indigenous. This native (indigenous) species is a widespread beach plant that grows as a low sprawling herb, sometimes in mats, and usually producing attractive rosettes. The alternate leaves are distributed along the stems or densely clustered toward the stem tips in whorls. They are thick but not fleshy, about an inch long, soft, silky, and silvery in color with dense hairs. The small white flowers with yellow centers have 5 or 6 petals, are fragrant, and are borne on 1 sided coiled spikes with branched stalks. This species is common in coastal sites on the older Islands, though said to be rare on Maui and Hawai'i. It also occurs throughout the islands of Polynesia.

The seaside heliotrope, *H. curassavicum*, is very similar, but does not form compact rosettes. The leaves of this indigenous plant are thick and juicy and not hairy, although they are also silvery in color. The flowers are similar but are found in more open, linear, less compact clusters. The flowers in these plants are borne on arms that

Look for: An attractive, small, silvery-leaved, rosette-forming beach plant with small white fragrant flowers.

remind me of the tentacles of an octopus. The spikes uncurl to reveal the new blossoms, which occur in pairs on one side of the spike in the position of the suckers on the arm of an octopus. I gather that this floral structure is typical of the heliotrope genus and some relatives. Seaside heliotrope is found along the sea coast on all the main Hawaiian Islands as well as in Australia, the Pacific islands, the southern U. S., South America, and the West Indies.

The leaves of both of these plants were used to make tea for medicinal purposes or as a tonic by the pre-contact Hawaiians. Hinahina was also used in preparations to treat thrush and asthma.

Look for: A similar but more sprawling beach plant with no rosettes and more open flower clusters.

'ILIMA
Sida fallax
Malvaceae (cotton, hibiscus, okra, milo)

One indigenous and 6 naturalized species in this genus are found in Hawai'i. The native plant above comes in two common forms. Along the beach and on arid lowland ridges it is usually a ground-hugging herb or low shrub, while in mesic to wet woodlands it may be seen as an erect shrub up to 4 or 5 feet tall. The alternate inch-long leaves are more or less pointed ovals with scalloped edges and are often velvety to the touch due to the fine hairs, especially on the lower surface of the leaf. The numerous pale orange-yellow flowers are borne individually on slender stalks that emerge from the base of a leaf. They are about an inch across and have 5 petals, with one side of each petal being longer than the other, giving the blossom a somewhat raggedy appearance. This plant is native to dry coasts and mesic forests from the Pacific islands to China. In Hawai'i it occurs on all the main Islands.

KEN SUZUKI

'Ilima is the island flower of O'ahu. It is probably the only pure-ly ornamental plant that was cultivated by the Hawaiians, as it was the premier lei flower. A lei often requires 1000 blossoms and at one time such lei were re-stricted to royalty. The Hawaiians also had a number of medici-nal uses for the plant. The bark from the tap root was part of a con-coction used to treat

Look for: A shrub or prostrate herb with 1 inch yellowish flowers with 5 petals, each of which has a lobe on one side that is longer than that on the other.

thrush. Flowers, leaf buds, and tap root bark were part of a mixture used for asthma. Sap from the mashed flowers provided a mild laxative for babies, and women chewed on the flowers to ease the pains of childbirth.

NEHE
Lipochaeta spp.
Asteraceae (lettuce, sunflower, silversword)

According to the *Manual* there were originally 18 endemic spe-cies in this genus in Hawai'i, but 11 of these are now endangered or extinct. These plants are some-what woody herbs with opposite leaves. The leaves of different spe-cies are highly variable in all other respects, however, some having leaves that are serrated ovals, or

Look for: A prostrate coastal herb with small oval hairy leaves and ½ inch yellow sunflower-like blossoms.

spindle-shaped, heart-shaped, intricately lobed and toothed, or even, as in one rare species, with branched filament-like leaves. All nehe have in common a small sunflower-like blossom with a yellow center and yellow petals around it. In some the center is large relative to the petals, and in others the reverse is the case. The plants inhabit a range of habitats in the Islands, although most probably favor drier areas. I will describe one of the commoner species, in the hopes that once you become familiar with this one, the others can also be recognized.

L. integrifolia is a common native shore plant. It is a prostrate hairy herb with numerous short, blunt somewhat fleshy leaves about an inch or less long. The leaves have no teeth and are oval or club-shaped with hairs on both sides. The stems tend to root where they contact the ground. Generally, a single flower is borne on a short stalk and is about ½ inch in diameter with about 8 petals. This coastal plant is found on all the major Islands.

PICKLEWEED
Batis maritima
Bataceae (no familiar relatives)

There is only 1 genus with 2 species in this family of plants, so it is not surprising that this introduced weed has no familiar relatives. The plant is a low, smooth, woody-stemmed shrub 2 to 3 feet tall with erect branches that grows in salty soil near the ocean. The opposite, succulent leaves are almost cylindrical and about 1 inch long. These contain a salty juice which smells like pickles. Inconspicuous green flowers and fruit are found in oblong cylindrical spikes at the base of each leaf. The plant is sometimes called ʻakulikuli kai because of a slight resemblance to the native ʻaku-likuli, but is more erect and lacks the attractive star-shaped flowers of the latter. Pickleweed is native to the tropical Americas and was first noticed in Hawaiʻi by Hillebrand in 1859 in the salt marshes of what is now called Sand Island on Oʻahu. It has spread over thousands of acres of sea-

Look for: A short, erect seaside weed with inch long opposite succulent leaves that emit a pickle-like salty juice when broken. Flowers and fruit are borne in warty cylindrical spikes at the base of each leaf.

side wetlands, displacing native marsh plants and disrupting nesting habitat for the Hawaiian black-necked stilt, on all the major Islands.

PORTULACA, 'IHI
Portulaca spp.
Portulacaceae (purslane, moss rose, miniature jade tree)

KEN SUZUKI

Seven species in this genus are found in Hawai'i, 1 indigenous, 3 endemic, and 3 introduced. They are generally prostrate succulent herbs of dry, usually coastal, areas. The plants you are most likely to see are both introduced weedy species. *P. pilosa* is a common volunteer in Honolulu gardens as well as on our dryland trails. It is a ground-hugging plant with reddish stems and small needle-like leaves that cluster along the stem. It can usually be recognized by the small (less than ½ inch diameter) magenta 5-petalled flowers that it bears. This pantropical plant was first collected on Kaua'i in 1922. It now occurs in dry and coastal habitats on all the major Islands.

Look for: A low growing sprawling plant with reddish stems, small needle-like leaves, and magenta flowers.

P. oleracea is a larger plant with coarse reddish stems and spatula-shaped leaves up to an inch long. It bears 5-petalled yellow flowers about ½ inch in diameter. This plant is also called "pigweed" or "purslane" and is eaten in salads in many parts of the world. Its origin is uncertain as it is now very widely spread throughout the world. It was probably introduced into Hawai'i as an ornamental or vegetable sometime before 1871. The plant is now common in low elevation disturbed areas on all the Islands.

Look for: A low, sprawling plant with coarse succulent reddish stems, spatula-shaped leaves, and small yellow 5-petalled flowers.

The Hawaiians pounded one of the native portulacas with bark from the mountain apple to obtain an extract that was used to sooth itchy skin and other skin disorders.

PUA KALA
Argemone glauca
Papaveraceae (poppy)

This native plant is an erect herb with alternate leaves, 1 to 4 feet tall, with yellow sap and generally very prickly on all parts except the flowers. The leaves are 3 to 8 inches long, clasp the stem, and are wavy, deeply lobed, and toothed. The plant has a bluish-green color due to a coating of wax that helps to reduce water loss in its arid habitat. It bears beautiful white six-petalled flowers with yellow centers. They are about 3 inches across and of typical poppy form. Small black seeds are produced in spiny oblong capsules. These endemics are found in dry coastal sites on the leeward side of all the main Islands and in dry subalpine areas on the higher ones.

Because this is one of the few native plants that is still well protected against browsing by its formidable prickles, several authors believe that it is probably a recent arrival in the Islands, perhaps reaching Hawai'i only a short time before the Polynesian settlers. In the absence of herbivorous animals, a plant which devotes its limited supplies of energy to reproduction will leave more descendants than one which spends part of them on the production of spines and repellant chemicals, so we predict that plants that have been in Hawai'i for a long time will lack these defenses. Unfortunately, of course, such plants are very vulnerable to herbivores when they do arrive, hence the imperiled state of so many of our natives. In any case, pua kala has apparently been in Hawai'i long enough to have evolved into a distinct species.

Little is known of the pharmacology of this poppy, although it apparently contains at least 3 narcotic compounds, including morphine. These are present in lower concentrations than in the opium poppy, and are accompanied by toxins which discouraged the Hawaiians from consuming preparations for their psychoactive effects. Extracts of crushed stems and grated roots of pua kala were applied to ease the pains of toothaches and sores, however.

JOHN HOOVER

Look for: An erect, bluish-green, usually very spiny herb with attractive 3-inch white poppy flowers and yellow sap.

SHRUBS

INDIAN FLEABANE, INDIAN PLUCHEA
Pluchea indica
Asteraceae (marigold, zinnia, lettuce)

These plants are erect coastal shrubs up to 6 feet tall with alternate stiff oval pointed leaves with coarse teeth on the outer half. The leaves are about 1 to 1 ½ inches long. Clusters of pinkish or purple flowers are borne at the branch tips in numerous small petal-

Look for: A compact coastal shrub with small, coarse, toothed leaves and clusters of small pink to purple petalless flowers.

less heads. The plant is native to southern Asia and was first found in Hawai'i in 1915. It is now widespread in low elevation dry coastal habitats on all the major Islands, where it may form dense thickets.

MA'O
Gossypium tomentosum
Malvaceae (hibiscus, milo, ilima, okra)

KEN SUZUKI

Look for: A sprawling shrub near the coast with gray-green, slightly downy leaves with maple-like pointed lobes, yellow blossoms, and small cotton balls.

Ma'o is an endemic cotton that occurs as a sprawling shrub on dry coastal plains and neighboring slopes. It has broad, gray-green leaves covered with fine white hairs. The leaves are 3 to 4 inches wide with 3 to 5 pointed lobes, somewhat like maple leaves in form. In season, the plant will have lovely bright yellow flowers about 3 inches in diameter, one of the more spectacular of our native blossoms. These are followed by small bolls that open to reveal a few seeds covered with short brownish cotton fibers.

The fibers of this native cotton are too short to be of commercial interest, but the ma'o crosses readily with its domesticated relatives and has been used in attempts to introduce genes for disease resistance and drought tolerance into those species. The native Polynesians extracted a yellow dye from the petals of the flowers, and a greenish one from the leaves.

NONI
Morinda citrifolia
Rubiaceae (coffee, gardenia, many natives)

Noni was introduced to Hawai'i by early Polynesian settlers who had many uses for the plant. It grows near the sea coast on humid, windward shores, and extends into the intermediate forest. The plant is a shrub or small tree with large glossy dark green leaves, that are often a foot or more in length. The leaves are elliptical, slightly pointed at the tip, and have prominent yellowish veins. The most distinctive feature of the plant is the fruit, however, which is actually a fused collection of many fruits. White, 5 petaled flowers emerge from a cluster near the branch tips. Later, these develop into a lumpy oval ball, divided into polygons representing the fused individual fruits. This ball, which may be as much as 5 inches long, turns yellow when soft and fully ripe. An uncommon endemic native, *M. trimera*, or noni kuahiwi, is found in certain areas of our dry to mesic forests. The leaves of this tree are quite different from that of the noni, but the fruit, although much smaller, is very similar in structure.

KEN SUZUKI

Look for: A shrub or small tree with large, glossy green leaves and a lemon-sized irregular lumpy fruit.

The Hawaiians ate the fruit of the noni in times of famine, but unlike some other Pacific Islanders, did not find it palatable when other foods were available. They obtained a yellow dye from the juice of the inner bark of the trunk or roots, and if this was treated with lime, a red or rose-colored dye was produced. These were used to color kapa cloth.

Noni was used in a great variety of medicines for many different ailments. The overripe, fermenting fruit has an odor like vomit, I am told, perhaps leading to a mental association with illness and medicine. Ripe fruit was used as a poultice for boils, and various parts of the plant, usually the fruit, were combined with a wide variety of other herbal concoctions and sea salt to treat asthma, menstrual cramps, sinus problems, chest pains, thrush, sprains, broken bones, pimples, kidney problems, constipation, and a host of other maladies. Noni fruit that had been mashed and cooked was even administered as an enema as part of a treatment for insomnia!

POHINAHINA, BEACH VITEX
Vitex rotundifolia
Verbenaceae (fiddlewood, lantana, Jamaica vervain, verbena)

This indigenous plant is a low, sprawling shrub that often roots where its nodes touch the ground. It may form extensive mats in coastal areas. The plant has opposite rounded oval leaves that are about 1 ½ inches long. The leaves are usually simple but are rarely divided into 3 leaflets. A crushed leaf has a distinctive odor of sage. They are gray-green above but paler and slightly fuzzy beneath. The flowers are clustered along the ends of the twigs and are about ½ inch long, a purplish-blue in color, and with 5 lobes, the lowest one being much larger than the other 4. This beach plant is native to the shores of the Pacific and Indian Oceans and occurs in Hawai'i along the coasts of all the major Islands.

Thomas H. Rau

Look for: A coastal plant with opposite gray-green rounded leaves that smell of sage when crushed and mint-like purplish-blue flowers with a prominent lower lobe and 4 small upper lobes.

TREES

HALA
Pandanus tectorius
Pandanaceae ('ie'ie)

For many years there was debate over whether hala was an indigenous plant that had reached Hawai'i by itself, or a Polynesian introduction. Then a cast of a hala fruit was found in an ancient lava flow on Kaua'i that was dated to one million years ago, so it seems that the tree has been here at least that long. The local trees may not have had all the desirable features the Hawaiians wanted, however, and they may have introduced other varieties from islands further south. Our hala species is found widely scattered in the Pacific islands from Java and the Philippines to Polynesia and the islands between. Hala is a tree of the mesic coastal environment, but is often

JOHN HOOVER

Look for: A spreading, branched, open tree with straight thick round prop roots and spiral tufts of long strap leaves crowded on the tips of the branches.

seen well up into the mountains to an altitude of about 2000 feet. It is a small tree with prickly, light brown bark that has marked rings where leaves were once attached. The ends of the branches carry spiral tufts of long strap-like leaves that clasp the stem and taper to a sharp point. Leaves may be 3 to 6 feet long and 2 to 4 inches wide. A cross section of the leaf has an "M" shape. Leaf edges and midribs bear sharp, hooked spines. Sexes are separate and the female tree bears large oval fruit, which resemble a pineapple in size and shape, and are sometimes passed off as pineapples by tour drivers who spoof the visitors. A distinguishing feature of hala is the number of thick straight prop roots that grow from trunk and branches toward the ground and help to support the tree.

Hala was a very valuable tree to the Polynesians. The keys from the fruit were used in lei, although only for ones own use—hala means "death" or "sin" as

13

well as being the name of the tree, so one would be malicious to give a hala lei to another. The seeds are edible and other portions of the fruit could be consumed in times of famine, although they were not very palatable and may have required special preparation. Old keys fray into tufts of fibers that were used as brushes in decorating kapa, and the other end of the key could be used as a stamp in the same process. It was the long leaves that were most useful, however. After the thorny edges were removed, long strips of a tough material were obtained that could be woven into canoe sails, mats, hats, bags, sandals, baskets, pillows and many other articles of use. In some areas the leaves were used to thatch houses. A preparation from the fruit was used to treat thrush, a yeast infection of the mouth in children that seems to have been common. The juice from the tip of aerial roots was a component in medicines to treat chest pain, vaginal discharge, and difficult child birth. It is rich in vitamin C and was used as a tonic. Young leaves were part of a mixture intended to cure dysentery.

Hala even plays a role in one Hawaiian origin myth. It seems that a goddess who was preparing hala strips cut her finger, and the clots from the drops of blood turned into two eggs, one male and one female. These then developed into the progenitors of the human race.

HAU
Hibiscus tiliaceus
Malvaceae (milo, okra, cotton, ʻilima)

We are not certain whether this tree is indigenous, having reached Hawaiʻi without human assistance, or was introduced by early Polynesian settlers because of the many uses they had for it. Hau is a large sprawling often prostrate tree that creates dense, nearly impenetrable tangles in low elevation valleys. No one who hikes in Hawaiʻi for long will be able to avoid acquaintance with this obstructive and exasperating plant. The

Look for: You can't avoid hau. This is the tree with rounded leaves that forms dense barriers of horizontal branches along and across trails in most of our low elevation valleys.

leaves of hau are broadly rounded and heart- shaped. They are 3 to 12 inches in each dimension and have a pale, slightly fuzzy undersurface. The yellow 5-petalled flowers are 2 or 3 inches across with deep red centers and fade to a burnt orange color as the day progresses. The tree is found around the world in the tropics and subtropics along coasts and in low-lying river valleys. In Hawai'i it is found along our wet coasts and in stream valleys on all the larger Islands.

Early Hawaiians obtained both course and finer fiber from the bark of hau, and these were used for a variety of purposes, from dragging canoe logs out of the forest to making fish nets, sandals, string figures, lei threads, slings, hula skirts, and bow strings. The poles were used for the outrigger booms of canoes, and short sections served as floats for fish nets. To make fire, a stick of hard wood, such as that of olomea, was rubbed vigorously against a dry piece of the soft hau until the resulting wood dust smoldered and could be used to ignite a tuft of tinder. Thin pieces of the light wood were used in kite frames. Flowers were consumed as a laxative, and the slimy sap was used for similar purposes or as part of concoctions to reduce congestion of the chest, ease labor pains, cure insomnia, and as an enema. Young leaf buds were chewed for dry throat. A hau branch was set erect in the ground ahead of the main body of the army before a battle, and as long as it stood, success in the fight was possible, but the side that was defeated let their branch fall.

MILO
Thespesia populnea
Malvaceae ('ilima, hibiscus, ma'o, abutilon)

This coastal tree may be indigenous or a Polynesian introduction. It has alternate, glossy, heart-shaped leaves 2 to 10 inches long, and 2 to 3 inch wide pale yellow flowers with purple centers. The fruit is a dry, flattened globe with 5 segments and about an inch in diameter. It is found along beaches throughout the Old World tropics. In Hawai'i it occurs on all of the main Islands.

The Hawaiians planted this tree around their homes for shade. The reddish-brown wood has a beautiful grain

Look for: An erect tree of Hawaiian beaches with glossy heart-shaped leaves and yellow hibiscus-like flowers.

and was carved into bowls for food and other uses. A yellow green dye could be obtained from the wall of the fruit. Occasionally, an exceptionally large straight tree may have furnished a log suitable for a canoe.

NAIO, BASTARD SANDALWOOD
Myoporum sandwicense
Myoporaceae (no familiar relatives known to me)

This plant may assume many forms, from a prostrate, sprawling shrub to an erect bush or fairly sizeable tree. On Oʻahu you are most likely to see it as a coastal shrub, but on Maui and the Big Island it is often a tree. The alternate leaves are crowded near the tips of the twigs and are spindle-shaped. They are about 2 to 6 inches long. Small white to pink 5-petalled flowers about ½ inch across cluster near the stem among the leaves. These are followed by whitish to purplish small round berries about ¼ inch in diameter. This indigenous tree is also found in the Cook Islands. In Hawaiʻi, it grows as a beach plant but extends upward through the dry, mesic, and wet forest to become one of the dominant vegetation forms in the alpine forest. It occurs on all of the larger Islands.

The wood of this tree has a pleasant fragrance much like that of sandalwood. When supplies of the latter were depleted during the days of the sandalwood trade, merchants attempted to sell naio wood to the Chinese instead, but this was rejected, possibly because the odor of these trees is short-lived, and so the derisive name "bastard sandalwood" was applied to them. The Hawaiians favored the hard wood of the plant for house frames, and because it burns so well, it was also used for fishing torches. Various parts of the plant were also used in medicinal mixtures to treat ailments ranging from asthma and serious lung infections with fever, to growths in the nose and preparations intended to ease childbirth.

JOHN HOOVER

Look for: A tree or shrub with slender oval pointed leaves clustered near the twig tips and ½- inch diameter, 5-petalled white to pink flowers on short stalks among them.

TREE HELIOTROPE
Tournefortia argentea
Boraginaceae (kou, comfrey, heliotropes, forget-me-nots)

This is a small umbrella-shaped tree, generally less than 20 feet tall, that is usually found very close to the sea shore. The leaves are alternate to whorled, large, 4 to 8 inches long by 2 to 4 inches wide, silvery-green, fleshy, fuzzy, and pointed oblongs in shape. Small white flowers are borne in pairs on one side of long, uncoiling, octopus-arm-like spikes, a typical heliotrope arrangement, also seen in hinahina. They are followed by small white fleshy fruit, about ⅜ inches in diameter. The tree seems to favor saline, nutrient-poor, sandy soils where it tolerates salt spray, strong sea winds, and intense sunlight. It is native to the shores of the Indian and South Pacific oceans, and was first collected in Hawai'i about 1864. It is an attractive addition to shoreline gardens and is now naturalized on all the major Islands.

KEN SUZUKI

Look for: A small tree with large oblong fleshy fuzzy leaves and small white flowers in pairs on uncoiling, octopus-arm-like spikes, that grows very near the shoreline.

Vines

BEACH MORNING GLORY, POHUEHUE
Ipomea pes-caprae
Convolvulaceae (morning glories, sweet potato, ung choi)

This indigenous beach plant has long trailing vines that root at the nodes and a thick tap root. The oval leaves, about 3 inches long, are often wider than they are long, and have a distinct notch at the tip. The species name is derived from the shape of this leaf, as "pes-caprae" means "goat's foot" which it was thought to resemble. The funnel-shaped flowers range in color from white through shades of pink and pale lavender. The 5-lobed blossom is about 2 inches in diameter and has a darker, often purple, throat that extends its star-like rays to the rim of the flower. This vine is found on beaches throughout the tropical world.

Early Hawaiian surfers would slap the surface of the ocean with these vines, believing that this would summon good surfing waves. In times of famine, the roots and stems of the plant were eaten, although in large amounts, they could be poisonous. The roots were included in a concoction designed to "purify the blood", and in another to treat lung troubles accompanied by high fever. Crushed leaves were mixed with salt and used as a dressing for sprains.

JOHN HOOVER

Look for: Long, sprawling beach vines that have large, broad leaves notched at the tips, and typical morning glory-type funnel-shaped flowers, in pink or lavender, with a darker star in the center.

KAUNA'OA, DODDER
Cuscuta sandwiciana
Cuscutaceae (dodders)
Cassytha filiformis
Lauraceae (cinnamon, camphor, avocado)

These two plants belong to entirely different families and are not at all closely related, but they are so similar in appearance that the Hawaiians gave them variations of the name "kauna'oa" and they can easily be described together for our purposes. Both are seen as networks of slender, thread-like, yellow-orange vines (*Cassytha* is sometimes green) with no leaves, but instead small suckers that penetrate the tissues of the plants they sprawl over or twine around and which extract nutrients from the host. Thus they are both parasites in which little or no photosynthesis takes place and in which the leaves have been reduced to tiny scales.

Cuscuta flowers are tiny compact whitish yellow, 5-petalled blossoms. Those of *Cassytha* are small and dull colored, with 3 parts. Both plants occur at low elevations, often near the coast. *Cuscuta* is an endemic native and favors other native plants as its hosts, although both are able to parasitize a wide range of plants including grasses. *Cuscuta* is found on all the major Hawaiian

Look for: Tangles of orangish threads draping other plants in lowland and coastal habitats.

Islands except Kaua'i and Kaho'olawe. *Cassytha* is indigenous to the Islands and is also found in coastal areas throughout the tropics. It occurs here on all the main Islands except Kaho'olawe.

Cuscuta may drain so much sap from its hosts that it kills them. It begins life as a seedling, rooted in the ground, but once it has established its parasitic way of life, it no longer needs a root and this disintegrates, leaving the plant with no connection to the soil. The early Hawaiians noticed this and called it the "motherless plant" since it had no attachment to mother earth! They made lei from the vines, and in fact, this plant was the Island Flower of Lāna'i. The filaments were mashed up with other plants to make a treatment for severe chest colds with phlegm. Another preparation of the mashed vines was given to women who had just borne a child in order to help expel the placenta.

PĀ'Ū-O-HI'IAKA
Jacquemontia ovalifolia
Convolvulaceae (morning glories, sweet potato, ung choi)

This indigenous vine grows near the coast in dry areas. It has small oval leaves, about 1 inch long, that have a notch at the tip. The small pale blue, 5-ribbed, bell-shaped flowers, less than an inch in diameter, are often mingled with the yellow blossoms of 'ilima to form a pleasing composition. The vines hug the

ground and may root at the nodes, allowing the plant to spread vegetatively. This species is also found in Africa, Mexico, and the West Indies.

Pā'ū means "skirt" in Hawaiian, and the plant name means "skirt of Hi'iaka". According to legend, when Hi'iaka was a baby, her older sister, the Volcano goddess Pele, left her on the beach while she went fishing. While she was gone, this plant grew

Look for: An attractive prostrate vine growing near the coast in dry areas with small oval leaves and light blue flowers.

over the child to shield her from the sun, which is how it got its name. The early Hawaiians mashed the leaves and stems of the vine to use as a laxative. They were also used in preparations to cure thrush in infants.

SEA BEAN
Mucuna gigantea
Fabaceae (Scotch broom, kudzu, mesquite)

These plants are large woody vines with bean-like compound leaves with 3 leaflets, each leaflet being a broad oval that comes to a point and about 2 to 5 inches long by 1 to 3 inches wide. The vine bears dangling clusters of sweet pea-like flowers of a yellowish green color. These are followed by flat pods some 4 to 6 inches long and 1 or 2 inches wide covered with very irritating orangish-brown hairs. It is believed that this plant is indigenous to Hawai'i, but it is also found in tropical Africa, Asia, and the islands of the Pacific. In Hawai'i it will generally be seen sprawling over rocks and trees near the ocean.

Look for: A large woody vine near the ocean with bean-like 3-part leaves and clusters of yellowish-green flowers or large flat pods.

The seeds of this plant were used in lei, and the seeds or the flesh of the pods could be used to produce medicines with violently purgative properties.

Dry Forest Plants

HERBS

AGERATINA, MAUI PĀMAKANI
Ageratina adenophora
Asteraceae (zinnia, sagebrush, sunflowers)

These weedy aliens are large, more or less erect, shrubby herbs that often have purple stems and an unpleasant odor. The opposite, 2 to 6 inch long leaves are oval or frequently diamond-shaped, but with the stem end of the diamond blunter and shorter than the tip end. The leaves have coarse teeth along the edges and 3 reddish veins spreading from the base. Flowers occur in crowded clusters on erect stalks. Each flower is about ⅛ of an inch long, lacks petals, and is fuzzy and white. The tiny seeds have small parachutes that permit wide dispersal by wind. "Pāmakani" means "windblown". The plant is common from dry to wet forests in Hawai'i. Its original home was in Mexico, but it was found here as early as 1885.

It is probably best to call this weed "ageratina" since there are at least 4 other plants, in 4 entirely different families, that are also called "pāmakani" in Hawaiian. A closely related plant, *A. riparia*, is also a common weed in Hawai'i. This is a sprawling herb with toothed, spindle-shaped leaves and flowers similar to those of the plant above. It is native to Mexico and the West Indies and has been observed in Hawai'i since 1926, where it occurs in dry to wet forest.

Look for: A coarse, shrubby herb with purple stems and bluntly, unsymmetrically diamond shaped leaves, that bears tight clusters of small fuzzy white flowers.

21

AGERATUM
Ageratum conyzoides
Asteraceae (wormwood, gaillardia, aster)

These weedy plants are odoriferous herbs up to about 3 feet tall, erect, but with rather weak, branching, stems and opposite hairy leaves. The leaves are broad, pointed ovals with small teeth along the edge and are usually widely spaced along the stem. They are 2 ½ to 3 inches long with a rough surface and prominent veins. Flat topped clusters of small purplish or white flowers, each about ⅓ inch across and lacking petals are borne at the end of the shoot. This plant is native to tropical America and was introduced into Hawai‘i, possibly as an ornamental, sometime before

Look for: A coarse, erect but flimsy herb with toothed opposite hairy leaves and small purplish (sometimes white) petalless flowers.

1871. It is now widespread on all the major Islands from sea level to about 4 or 5,000 feet of elevation, in dry to wet habitats.

AIR PLANT
Bryophyllum pinnatum
Crassulaceae (donkey tail, jade plant, hen and chickens, kalanchoe)

This alien plant is an erect herb, usually unbranched, that is a couple of feet tall, rarely reaching 5 feet or more. It has somewhat fleshy leaves that are opposite, oval in shape, and 2 to 6 inches long by 1 to 3 inches wide with scalloped edges. At top and bottom of the hollow stalk the leaves are usually simple, but in the middle they are often pinnately compound, with 3 to 5 leaflets arranged along the central stem. The flowers are pale yellow cylinders tinged with red and dangle from the stem. They are an inch or a bit more long, inflated and papery. Before they open, they are like small balloons

KEN SUZUKI

22

and will burst if squeezed. This plant is now so widely scattered throughout the tropical and subtropical world that we are not sure where it originated. It was established in Hawai'i in the district of Ka'ū on the Big Island before 1871, however, and is now common in low elevation dry and mesic areas on all the major Islands.

Stems of this plant will root where they touch the ground. Small plantlets also develop in the notches along the edge of the leaf, even if the leaf is hung on a wall, which is why it is called "air plant". This exuberant vegetative reproduction probably explains its wide distribution, although it is also occasionally grown as a garden curiosity.

Look for: An erect herb with opposite scalloped oval leaves and papery cylindrical dangling flowers.

AMARANTH
Amaranthus spp.
Amaranthaceae (pig weed, celosia, pāpala, kuluʻī)

There are 6 species of amaranth in Hawai'i—one rare endemic found only on Ni'ihau which you are extremely unlikely to see, and 5 introduced weeds. Once you are familiar with one of the weedy species you will probably have no difficulty in recognizing the rest, so I will only discuss the one that is most likely to impose itself on your attention—the spiny amaranth. This plant, *A. spinosus*, is a vigorous herb growing from 1 to 5 feet tall with many branches. The stem may be green or red. The leaves are oval to nearly tri-angular in shape, 1 to 5 inches long by 1 to 2 ½ wide. A pair of spines each of which may be as much as an inch long, emerges from the stem at the point of leaf attachment. Tiny green flowers are tightly clustered in long, branched or unbranched spikes that are found at the base of leaves or on top of the plant. This plant is native to many of the warmer parts of the world and is now widespread in Hawai'i in low elevation, disturbed, dry to mesic sites. It was probably introduced to the Islands by accident before 1928. The other common species are similar to this one but lack spines.

Look for: A hardy branched herb, up to 5 feet tall, with coarse oval to triangular leaves, and spikes of tiny, densely clustered flowers or seeds, in dry disturbed areas.

Several amaranth species are used in

India and other parts of the world as spinach greens. Others were domesticated for their seeds, which were one of the major food grains of the Aztecs in Mexico and are also an important grain in Ethiopia. This is one of the few plants that was independently domesticated for food in both the Old and the New Worlds. Today, you are most likely to be familiar with its relatives in their role as garden weeds or ornamentals.

CARRION FLOWER
Stapelia gigantea
Asclepiadaceae (milkweed, crown flower, pakalana)

Some people might mistake this plant for a kind of cactus. Round stems of the herb creep along the ground and send up erect, succulent stalks, 5 or 6 inches high, each with 4 prominent ribs that bear soft curved points along each edge. These stalks have no leaves or thorns but are soft and slightly fuzzy. Large 5 pointed star-shaped flowers, 4 to 6 inches in diameter, with a pattern of dull reddish wavy lines against a pale yellowish background are produced on short stems from the base of the plant. These flowers often have a fragrance like decaying meat, and attract flies which may pollinate them. Seeds are produced in paired finger-like pods. Carrion flower is native to southern Africa and was probably brought to Hawai'i before 1871 as a garden plant. It is now common in the dry lowlands on O'ahu, at least.

KEN SUZUKI

Look for: Clusters of low growing, 4 ribbed, succulent stalks in dry areas on O'ahu that often have large star-shaped flowers with an odor like decaying meat.

This plant is also called giant toad flower, starfish flower, or Zulu giant. Given the distinctive and identifying odor of many specimens, however, carrion flower seems to be the most suitable name for it. I have grown this plant in my yard for many years, and once noticed some tiny maggots crawling around in one of the blossoms, in a state of desperate frustration, I presume, in search of the carrion which the mother fly must have believed was present when she laid the eggs.

COAT BUTTONS
Tridax procumbens
Asteraceae (fleabane, dandelion, aster)

This common, low-elevation weed is a sprawling herb with a tap root, that has opposite, arrow-head shaped leaves with prominent teeth, especially near the base. All parts of the plant are hairy. The rounded flower head is borne on a long erect fuzzy spindly stalk, and opens into a small daisy-like blossom about ¾ inch in diameter with short pale yellow petals. The tiny seeds are spread by the wind on small fuzzy parachutes. The plant is native to South and Central America but is now widespread in the tropics. In Hawai'i it is common in low dry disturbed habitats on all the main Islands. It was first collected on Maui in 1922.

This is a common weed in the drier parts of Honolulu where I live. It is difficult to pull up because of its tendency to break off at the base, leaving the root in the ground.

Look for: A sprawling, hairy herb with prominently toothed leaves and long spindly flower stalks bearing small daisy-like flowers.

COMMON BASIL, SWEET BASIL
Ocimum basilicum
Lamiaceae (mint, basil, selfheal, sage)

These introduced plants are branching herbs with opposite leaves. The hairy leaves are oval with pointed ends, prominent veins, and often small teeth along the edges. Twigs tend to have 4 angles, like many mints. The small white flowers occur in slender spikes and often have a pink or lavender tinge. Probably the most reliable way to identify the plant is to crush a leaf and enjoy the typical basil fragrance. This plant is well known for its role in Italian cooking and was probably introduced to Hawai'i for

Look for: A bushy herb of dry lowlands with slender flower spikes and squarish twigs whose crushed leaves have the odor of basil.

25

culinary purposes. The original home of this herb is uncertain as it is now found throughout the tropical world. It was first reported here in 1912 and is naturalized in dry, lowland habitats on all the major Islands.

Other members of this genus are also naturalized in the Islands. Some of these are used in various Asian cuisines. They may look somewhat different from this plant, but all will emit a strong basil-like odor from the crushed leaves or fruit.

CORAL BERRY
Rivina humilis
Phytolaccaceae (pokeweed)

This plant is an erect, branching herb about 1 or 2 feet tall that often forms dense masses over considerable areas in dry and mesic forest. It has alternate pointed oval leaves about 2 to 4 inches long. Small white propeller-shaped flowers with 4 petals are tightly clustered along 3 to 6 inch long spikes and these are followed, often at the base of the same spike, by small red berries. These colorful spikes are the most conspicuous identifying feature of the herb. The plant is native to the southern U. S. and South America, and apparently reached Hawaiʻi, probably as an ornamental, before 1871.

JOHN HOOVER

Look for: A low growing herb of dry and mesic forest, often in shaded areas, with conspicuous spikes of white, 4-petalled flowers and/or small red berries.

The juice from the berries has been used as an ink or dye. The roots and berries may be slightly toxic and should not be eaten.

DESMODIUM, SPANISH CLOVER
Desmodium spp.
Fabaceae (koa, klu, monkey pod, and all the great bean family)

There are 7 species of desmodium in Hawaiʻi, all of them inconspicuous but noxious little alien weeds. Most of our species are native to tropical or subtropical America. They were introduced at various times, the most common species having been observed here by 1840. They are sprawling, low herbs

with compound leaves that have 3 leaflets, each about an inch long. Some have a light colored streak down the center of the leaflet. The flowers are tiny pea-like blossoms with a lovely pink to purple color. The fruit is the most distinctive part of the plant, however, and the one you will find it impossible to ignore, as the chains of half-moon-shaped seeds cling to sox, pants, and packs, and require much time and effort to remove when they are numerous, as

Look for: A pernicious low, scraggly herb bearing chains of half-moon-shaped seeds that stick to clothing.

they often are. A friend of mine calls these "velcros", which is quite appropriate. There is some variation in the shape of the seed in different species. This pest is a plant of dry and intermediate forests, but is willing to grow almost anywhere when given a chance. Just within the last 3 or 4 years it has become much more widespread and is now found on trails everywhere on Oʻahu.

ʻENAʻENA
Gnaphalium spp.
Asteraceae (daisy, silversword, coreopsis)

Of the 200 or so species in this genus, 3 are found in Hawaiʻi, 1 endemic and 2 as accidental weedy introductions. All are low growing, woolly, silvery-gray herbs that form small rosettes 2 or 3 inches in diameter, in open spaces. The leaves are paddle-shaped, broader near the tip than at the base, and without teeth. The flower heads are clustered at the top of a stalk that is usually less than 1 foot tall. Flowers are brownish to yellowish in color, and not very conspicuous. One of the alien species is native to Australia and the other to North America. They were introduced into Hawaiʻi in the late 19th and early 20th centuries, and are now found in dry to wet open areas. The endemic species is gener-

Look for: A low growing, 2 or 3 inch wide rosette of silvery-gray leaves in open areas where it is not shaded out by taller vegetation.

27

ally found in dry areas near sea level or on lava and cinders at higher elevations. These plants are sometimes called "cudweed" or "everlasting", the latter name referring to the durable nature of the dried flower stalk.

The Hawaiians placed ʻenaʻena among their feather kahili during storage to repel insects, a property that may have been due to the presence of aromatic volatile oils in the plant. "ʻEnaʻena" means "hot" which may have referred to this repellant quality.

FLORA'S PAINTBRUSH
Emilia spp.
Asteraceae (thistle, dandelion, daisy, sunflower)

There are 3 species in this genus in Hawaiʻi. They are low growing herbs with a cluster of leaves at the base and alternate leaves further up the stalk. The basal leaves are 2 to 4 inches long, paddle-shaped or with wavy teeth or lobes. One or a few flower heads form at the tips of long slender upright stalks a foot or so tall. Each flower is a vase-shaped structure about ½ inch long with a spreading tuft of orange, red, or purple filaments forming a blossom about ½ inch or less in diameter on top. The fruit consist of tiny, parachute-equipped seeds in a ball like that of the dandelion, but smaller. The plants are native to the Old World tropics but are now widely distributed weeds. They were first collected in Hawaiʻi in 1895 and are now naturalized in dry disturbed habitats on all the major Islands.

Look for: A low growing herb of dry areas forming a rosette with a tall, slender stalk bearing one or more reddish, paintbrush-like blossoms.

HAIRY HORSEWEED
Conyza bonariensis
Asteraceae (dandelion, goldenrod, lettuce)

This plant is an erect hairy herb, up to 3 or 4 feet tall, with gray-green, slender leaves. The leaves are about 2 to 6 inches long by ¾ inch wide and the edges

Look for: An erect hairy gray green
herb with slender pointed leaves that
are edged in small steps.

have small, abrupt steps along each side. In
season, the weed produces numerous small
fuzzy white flowers. The plant may have
come originally from South America, but
is so widespread that its homeland can not
be assigned with certainty. In Hawai'i it is
often found in dry areas, but may extend to
mesic or even wet forests. It was apparently naturalized here before 1871.

A very similar plant, horseweed, *C. canadensis*, with somewhat lighter-col-
ored leaves is also common in Hawai'i.

'ILIE'E, PLUMBAGO
Plumbago zeylanica
Plumbaginaceae (limonium, statice, plumbago)

This indigenous plant is a
sprawling, low shrub with alternate, broadly oval pointed
leaves 1 to 4 inches long by an
inch or two wide. The most
distinguishing features of
'ilie'e are the flowers, each of
which emerges from an oval
sheath about ½ inch long
that is covered with coarse
sticky hairs. These structures
are borne on the sides and tip
of a slender, linear stalk. Five
white petals expand from the
end of a floral tube about ½
inch long. The plant is found

Look for: A low growing plant of dryland
areas with numerous small 5-petalled white
flowers emerging from oval sheaths covered
with coarse, sticky hairs.

widely dispersed through the Old World tropics. In Hawai'i, it is found in
dry lowland areas and dry forests and shrublands.

The leaves of 'ilie'e were mashed up in a mixture to make an ointment used on
sprains. Juice from the roots of the plant was used as the pigment in tattooing
in both Hawai'i and Africa.

KOʻOKOʻOLAU
Bidens spp.
Asteraceae (sunflowers, asters, daisies, silverswords)

JOHN HOOVER

There are 23 species of this genus in Hawaiʻi, 4 alien weeds and 19 endemic natives. These plants are herbs or small shrubs with light green simple or compound leaves. The ones with compound leaves often have leaves that look like those of roses and have small teeth along the edges. The flowers are the most distinctive fea-

Look for: A 2 to 4 foot plant with a scattering of small sunflower-like yellow blossoms, or white blossoms and beggar tick seeds.

ture of the plants, looking like miniature sunflowers, less than an inch to 2 inches across, and often having gaps among the petals. All of the natives have yellow flowers, but at least one of the common introduced species has white flowers and another lacks petals altogether. The alien weeds produce seeds with bristles that bear tiny barbs allowing them to cling to fur, feathers, or clothing as a way to disperse. These seeds are commonly known as beggar's ticks or Spanish needles. Most of the native species have lost their barbs and do not collect on your clothes. This is a large genus with about 230 species world wide, mostly in North and South America, Africa, and Polynesia, but with some in Europe and Asia as well. In Hawaiʻi, different species grow from the coast through dry and mesic forest and well into the wet forest.

The Hawaiians made a tea from the leaves of this plant, with different species producing teas of subtly different flavors. This was used as a general tonic and to accompany a great variety of treatments for a galaxy of ailments. It was also used as a douche to clean the vagina when a woman had had several miscarriages, and added to the diet of a pregnant woman to strengthen the body of the child. After childbirth, it was thought to aid recovery and stimulate milk production. This herbal tea is still available in health food stores.

LION'S EAR
Leonotis nepetifolia
Lamiaceae (lavender, mints, oregano, salvia, horehound)

This introduced weed is a coarse, erect herb that grows up to about 6 feet tall. It has opposite, rounded-triangular leaves with teeth along the edges. As in many members of the mint family, the stems are square in cross section.

The small flowers are a bright orange color. Flowers and fruit occur in dense round spiky balls that are arranged at well-spaced intervals along the stem, like so many small sea urchins on a kebab skewer. Each ball is about 1 to 2 inches in diameter. The plants are sometimes used in dried flower arrangements. This weed, originally from South Africa, was probably introduced as an ornamental and was first collected on Oʻahu about 1938. It is now widespread in dry lowland areas on all the larger Islands.

KEN SUZUKI

Several other weeds in closely related genera, including members of *Leonurus* (lion's head), *Marrubium* (horehound), and *Mentha* (mint), have been introduced into Hawaiʻi. These also have the flowers and fruit clustered at intervals along the stem, but in these plants the fruiting bodies are smaller and less spiky.

Look for: A weedy dryland plant with large round spiky balls well spaced along an erect stem.

MOLASSES GRASS
Melinis minutiflora
Poaceae (grasses, bamboo, sugar cane, and all the major food grains)

This is an erect or often sprawling grass of dry, mesic and even fairly wet, disturbed habitats. It has a sweet, molasses-like odor and the light gray- or blue-green leaves are sticky and very hairy. The plant comes from Africa and was introduced to Hawaiʻi about 1914 as cattle fodder. It has become a serious pest in dry areas, forming dense mats 2 or 3 feet thick, covering and choking out native vegetation and preventing the establishment of seedlings. It is adapted to burning and helps to support fires that threaten our native dry forest, while recovering rapidly after a fire, itself. In the past, there was a relatively sparse ground cover in our dry forest, so that the infrequent fires were unable to spread into it very readily. Now with this grass and other alien plants providing a much denser cover in dry areas, combined with many more fires due to human activities, the native dry forest is exposed to more frequent burning, and many interesting native plants are being eliminated from this forest zone as a consequence.

The flowering head of molasses grass has a reddish tinge, like Natal red-top, and in season, (early winter) meadows of the plant form an attractive sea of waving pink plumes. I still remember what a beautiful picture the wine-colored tufts of this grass made one day against the yellow stalks of a dry meadow with scattered clumps of bright green fern, the blue sky, and a few fluffy white clouds, as I was hiking on a plateau on the lower slopes of the Koʻolau Range overlooking Pearl Harbor, the Central Valley, and the distant Waiʻanaes.

Look for: Dense mats of a light green, very hairy grass that smells sweet and is sticky, in dry disturbed areas.

PARTRIDGE PEA
Chamaecrista nictitans
Fabaceae (clover, beans, wiliwili, mamane)

This introduced plant is a common erect herb 2 to 3 feet tall, with alternate pinnately-compound leaves that have 15 to 30 pairs of small oval leaflets arranged along the stem. Small yellow flowers, about ½ inch across appear on the stalk near the base of a leaf and these develop into 1 or 2 inch long linear bean pods containing rows of small brown beans. The plant is native to the American tropics and occurs in Hawaiʻi in dry or mesic disturbed sites. It was first reported here before 1871. The vegetative parts of this plant are similar to those of the sensitive plant, *Mimosa pudica,* but unlike that plant, its leaves do not fold up when touched.

Look for: A common erect herb of dry areas with alternate compound leaves and small yellow flowers. Leaves of this plant do not fold up when touched.

PŌPOLO, BLACK NIGHTSHADE
Solanum americanum
Solanaceae (nightshade, potato, jimson weed, datura, petunia)

This plant may be indigenous, but we are not sure. It is an herb up to 4 feet tall with oval leaves, 1 to 4 inches long, that taper to a sharp point. The edges are often irregular. The shrubby herb bears clusters of small star-shaped ¼ inch wide white or purple-tinged flowers. These are followed by small black juicy berries. The plant is widely distributed throughout the world in tropical and warm temperate climates. In Hawai'i it is found in coastal to wet forests and subalpine woodland on all the main Islands.

There is some question about the toxicity of this plant. Some authors admit that the ripe berries are edible, but caution that unripe berries and greens may be poisonous. Young shoots are cooked and eaten on other Pacific islands, however, and both berries and greens were apparently eaten in Hawai'i when food was scarce, and young leaves were eaten raw with a meal to prevent bloating. The berries were also used in making a purple-black dye for kapa. Pōpolo was one of the

Look for: A large, open herb with clusters of small black berries.

most important medicinal plants to the Hawaiians. Preparations of the plant were used to treat coughs, or made into a tea to serve as a tonic. Mashed leaves were applied to inflamed eyes and mixed with salt to apply to wounds. Juice from the leaves was rubbed into sore muscles, tendons, or joints. The plant was considered to be one of the embodiments of Kāne, and concoctions including it were considered efficacious for all respiratory problems, skin eruptions, cuts, and wounds. Young leaves steeped with salt were taken to tone up the digestive system. Pōpolo was also used to treat thrush, vaginal discharges, sore throat, and tuberculosis.

SENSITIVE PLANT, SLEEPING GRASS
Mimosa pudica
Fabaceae (kiawe, soybean, indigo, koa)

This small introduced weed is quite unmistakable. It is a small erect or creeping herb, sometimes spreading as much as 2 or 3 feet, though usually smaller, with pinnately-compound leaves that have 10 to 26 pairs of small oval leaflets along each stemlet. The flowers are small pink fuzzy balls about ½ inch in diameter, followed by clusters of 2 to 8 pods containing pale brown

Look for: A low creeping herb with purplish stems, pink puff-ball flowers, and leaves that fold up and droop when touched.

seeds. The stems are purplish in color and bear small prickles. If you touch the leaves, they quickly droop and fold up, the younger ones responding more rapidly than the older. After a time, they will slowly unfold again. They also fold up at night. This plant is a pantropical weed, that probably had its origin in South America. It is naturalized in Hawai'i in open dry to wet disturbed areas. It was first collected on O'ahu in 1864 or 5, but was already widespread by that time.

SPURFLOWER, PLECTRANTHUS, 'ALA'ALA WAI NUI PUA KI
Plectranthus parviflorus
Lamiaceae (mint, salvia, rosemary, thyme)

This indigenous plant is another of our native mints, but so different in both appearance and habitat that I have chosen to consider it separately. Plectran-thus is an erect but low-growing herb with opposite leaves that are roughly rounded triangles in shape, fuzzy, and with rounded teeth on each edge. They are typically about an inch long. Like our other native mints, the plant is not aromatic. Flowers are borne on slender erect spikes and are small—less than ¼ inch long. The blossoms are white or pale blue with the usual bilateral symmetry of the mint family. In this species the lower lip is larger than the upper, resembling the native *Phyllostegia* in this respect. The flowers are followed by tiny brown seeds. This native is also found in Australia and adjacent islands and in Polynesia. In Hawai'i it occurs in dry, rocky, exposed lowland sites on all the main Islands.

Look for: A low-growing plant of exposed, dry, rocky areas with opposite fuzzy roughly triangular leaves edged with rounded teeth.

I have found no reference to any use of this plant by the original Hawaiian inhabitants.

'UHALOA
Waltheria indica
Sterculiaceae (cocoa, cola nuts - soft drink flavoring)

This indigenous plant is found throughout the tropics. In Hawai'i it is typical of dry, disturbed areas. The plant is a small shrub or woody herb, usually 1 to 3 feet tall and may be erect or sprawling. The alternate leaves are gray-green pointed ovals about 1 to 6 inches long by ½ to 2 inches wide with small teeth along the edges. They have a velvety texture thanks to a dense coating of hairs, and prominent veins, indented on the top surface and protruding from the lower. Furry little irregular balls occur in the angle between the stem and the leaf stalk and minute 5-petalled yellow flowers appear among these.

KEN SUZUKI

The Hawaiians chewed the bitter root bark to relieve sore throats, or prepared a gargle from pounded stems, leaves, or root bark for the

Look for: A common, weedy plant of dry, open areas with velvety gray-green toothed leaves with prominent veins.

same purpose. Various parts of the plant were used as components of medicines used to treat thrush in children, chest pain and congestion, asthma, as a blood cleanser, and in a douche for women who had suffered several miscarriages.

VERBENA
Verbena litoralis
Verbenaceae (vitex, Jamaica vervain, lantana, fiddlewood)

This naturalized weed is an erect herb from 1 to 6 feet tall with quadrangular stems and opposite oblong leaves about 1 to 4 inches long by ½ inch wide. The leaves have prominent saw teeth along each edge and each pair emerges at right angles to the pairs above and below it. Branches and flower spikes often emerge in sets of 3 from their origin at the end of a twig. Many long slender spikes emerge from the branch tips or leaf axils and these bear tiny blue flowers with

5 lobes, which tend to occur in a ring around the enlarged oval tip of the spike. The plant is native from Mexico to South America and was first collected in Hawai'i in 1837. It is now common in dry to wet forest on all the main Islands.

The juice of the mashed plant was used to treat skin ailments, cuts, and bruises and even for sprains and fractured limbs by residents of Hawai'i after its introduction into the Islands.

Look for: An erect open weed with square stems, opposite, saw-toothed leaves, and long spikes bearing tiny 5-lobed blue flowers.

SHRUBS

'A'ALI'I
Dodonaea viscosa
Sapindaceae (lonomea, lychee, longan, ferntree)

This plant is a shrub or small tree with slender twigs and wavy, papery, alternate leaves. The leaves are 2 to 4 inches long, spindle-shaped, with the widest part in the middle or toward the tip, and, especially when young, with a sticky or resinous coating. The two sexes tend to be on separate plants. The male flowers are inconspicuous cups with 10 stamens. The female plant bears attractive "paper lantern" fruits about ½ inch in diameter that vary in color from pale

KEN SUZUKI

Look for: An open shrub or small tree with alternate papery pointed leaves and, in season, attractive "paper lantern" winged capsules in cream to pink or maroon colors.

greenish-cream through shades of pink to a deep maroon. Each has 2 to 4 prominent semi-circular papery wings. This indigenous shrub is found all over the tropical world and is one of the few plants that are native to both Hawai'i and the Mainland U. S., occurring in Florida and Arizona as well as on the Islands. 'A'ali'i is a very hardy plant, adaptable to a wide variety of habitats. In Hawai'i it is most abundant in open dry areas, but extends from the coast through dry, mesic and wet forest into subalpine shrublands, and is found on all of the major Islands.

The fruit and leaves of 'a'ali'i were used in lei by the Hawaiians and a red dye for kapa was obtained from the deeply colored fruit. The hard, yellow-brown wood of the plant is very durable and they used it for 'ō'ō, or digging sticks, spears, tools, and house posts. The shrub is unusual among Hawaiian plants in having a tap root, and is very resistant to the wind. Natives of the Ka'ū area on the Big Island used to boast, "I am an 'a'ali'i; no wind can push me over." 'A'ali'i is one of the few Hawaiian natives that is fire adapted. It is often able to resprout from the base of a burned plant, and the germination of the seeds may be stimulated by exposure to smoke or heat. The plant is a host of the beautifully colored koa bug and also of the small bluish Blackburn butterfly, *Udara blackburni,* one of only two butterflies native to Hawai'i.

ABUTILON
Abutilon spp.
Malvaceae (milo, cotton, ʻilima, hibiscus, okra)

There are 3 rare endemic species in this genus, 3 introduced weeds, and 1 that is questionably indigenous. These plants are generally medium-sized or small shrubs with hairy, heart-shaped leaves and small 5-petalled hibiscus-like flowers. *A. grandifolium*, one of the introduced species, is the one you are most likely to encounter, and so I will describe it. This plant is a shrub up to 6 feet tall. The hairy heart-shaped leaves are as much as 6 inches long, and the basal lobes tend to overlap. The orange flowers are about 1 inch across and resemble those of the native ʻilima. They are often used in lei in place of ʻilima flowers. The fruit is a flattened, deeply ribbed, pumpkin-shaped body about ½ to 1 inch in diameter. The plant is native to the New World tropics and was first collected in Hawaiʻi in 1903, possibly having been introduced as an ornamental. These plants are now fairly common in dry lowland sites on all the main Islands.

A. menziesii, or koʻoloa ʻula, is a rare endemic native of dry forests in Hawaiʻi. This attractive plant has pale silvery-green, hairy, heart-shaped leaves and pendulous flowers about 1 inch in diameter that range from pink to maroon in color. It has been used to a limited extent in home gardens and landscaping and deserves more attention in this respect.

KEN SUZUKI

Look for: A dryland shrub with hairy heart-shaped leaves whose basal lobes overlap, with small orange ʻilima-like flowers, and small flattened pumpkin-shaped fruit.

ʻĀHEAHEA
Chenopodium oahuense
Chenopodiaceae (beet, spinach, quinoa, pigweed)

This endemic shrub is small to medium sized, with alternate gray-green leaves that are silvery underneath. The leaves have an unusual shape that is very distinctive once seen, but difficult to describe. They are roughly triangular in shape, with rounded corners and irregular, wavy edges and 3 or more shallow lobes. Leaves are about 1 to 1 ½ inches broad and long. I am told that they closely resemble a weed called "lambs quarters" or "pigweed", which is a related chenopodium. The flowers are small greenish balls that occur in open clusters at the end of a stem and mature into a similarly shaped dry brown fruit. Some plants have a distinct unpleasant ammoniacal or "fishy"

odor, which, when present, is a major distinguishing characteristic. This native plant is found near the coast or in dry forest up to subalpine shrubland, on all the major Islands and on atolls of the northwest Hawaiian chain.

Pre-contact Hawaiians cooked young leaves of ʻāheahea in an imu and ate them in times of dearth. The strong wood of the shrub was used for hooks to catch sharks and other fish.

Look for: A dryland shrub with alternate gray-green oddly shaped leaves and sometimes an unpleasant odor.

ʻĀKIA
Wikstroemia spp.
Thymelaeaceae (I know of no familiar relatives)

The ʻākia are shrubs or small trees. One, *W. uva-ursi*, is now widely used in landscaping in the Islands, especially in dry areas. Look for it in the xeriscape gardens of the Honolulu Board of Water Supply. This species is a low, sprawling shrub with rounded leaves, that is native to coastal areas. The most common Oʻahu species is found in the dry forest and above, and seems to be one of the few native plants that can compete successfully with the numerous introduced weeds in the drier areas. It is a shrub or small tree with pale (sometimes dark) green leaves that lie opposite each other along the twig, are 1 to 3 inches long, and are pointed, unlike the more rounded leaves of ʻakoko. The bark is smooth, and reminds me somewhat of cherry bark, often with a reddish or purple-red tinge, especially on younger branches. There are often light bands or blotches running across more mature branches. Flowers are small with 4 petals and greenish-yellow in color. The fruit is oval, about ½ inch long, and orange to red. The sap is not milky. This is a very variable species, so no trait mentioned above will necessarily be true of every plant. The 12 species in this genus are all endemic to Hawai‘i.

Thomas H. Rau

The strong fibers from the bark of this plant were sometimes used as cordage or rope by the Hawaiians. Its leaves were included in a preparation used to treat stubborn cases of asthma. Many continental relatives of the ‘ākia are notorious for their poisonous character. At least one species here (we're not sure which) was used to stun fish. Roots and bark of the plant were pounded and mashed and then scattered through a tide pool. After a few minutes, small fish would float to the surface, where

Look for: A shrub with spindle-shaped opposite leaves on slender twigs with smooth bark that often has a reddish tint. The sap is not milky.

they could be gathered. Any fish that escaped notice would recover as the poison was diluted out by wave action. It has been reported that an infusion of the roots and bark of a species of ‘ākia was part of the cocktail administered to criminals to execute them in old Hawai‘i. While some doubt has been cast on the idea that Hawaiian ‘ākia berries are poisonous, I would not recommend including them in a snack.

‘AKOKO
Chamaesyce spp.
Euphorbiaceae (spurges, poinsettia, castor bean, kukui)

There are many ‘akoko in the Islands, 22 species in this genus according to the *Manual*, 7 introduced (including a variety of garden weeds) and the rest endemic. They are quite diverse in appearance, and occur in all zones, but can generally be recognized as belonging to the same family. The ‘akoko you are most likely to encounter along the trail is a shrub of the dry forest. It has small, rounded, slightly gray-green or blueish-green leaves ½ to 2 inches long that are arranged opposite to each other along the twig and often lie in a flat plane. This is a significant feature. If you look at other plants that have opposite leaves, such as ‘ākia or strawberry guava, each successive pair of leaves emerges from the twig at an angle, usually near 90 degrees, relative to the previous pair. The leaf stem may then bend to orient the leaf toward the light. Very few plants have successive pairs of leaves all emerging in the same

39

plane. The twig of the ʻakoko is slightly reddish with lighter bands running across it and has a knobby appearance. Another major distinguishing feature of the ʻakoko, is seen if you remove a leaf from the plant. A bead of milky white fluid oozes out of the twig where the leaf was removed, and another from the broken leaf stem. This will allow you to distinguish this plant from the rather similar ʻākia, which also occurs in this zone, but does not have milky sap. If you look closely, you may notice the fruiting body, which looks like a tiny cup with a bladder attached to its top. In the commonest species, *Chamaesyce celestroides*, this little three-lobed bladder stands straight up from the middle of the cup, but in most species with such small fruits, the capsule dangles at an angle on a short stalk from the cup. "Koko" means "blood" in Hawaiian, and "ʻakoko" is "blood colored", perhaps referring to the color of the ripe fruit.

JOHN HOOVER

The ʻakokos are host to an interesting native planthopper, *Dictyophorodelphax*. This little insect has a snout as long as the rest of its body, and a loop of the gut runs up into it. It seems likely that this extra-long digestive organ is required to process the latex-like sap of the plant and remove any toxins present, which these milky saps often contain, so that the animal can digest it. Or maybe the snout just makes the insect look like a twig or thorn on the plant so that it will escape the notice of a hungry bird.

Look for: A shrub or small tree with gray-green opposite leaves in a plane, and milky sap.

The sap of the ʻakoko may have been used as an ingredient in the paint that was used to paint canoes black, although the juice from the inner bark of the roots of kukui trees was probably preferred. The bark, or sap, was also used in two different preparations with the sap of a green kukui nut to make potent purgative medications.

BURR BUSH
Triumfetta spp
Tiliaceae (linden, white moho)

There are two species of this pernicious alien weed in Hawaiʻi, both being herbs to shrubs up to 6 feet tall and characterized by small round burrs, about ¼ inch in diameter, that cling to clothing and packs and are thus spread along our trails. Both have coarse broad alternate serrated leaves of somewhat irregular shape, often tending to have 3 to 5 shallow lobes with a major mid-rib in each. Stems and leaves are hairy. Leaves are often badly damaged by insects. The flowers are yellow with 5 petals and about ⅓ inch in diameter. The burrs cluster near the ends of long slender twigs. The leaves when crushed are said to have a peculiar odor. *T. rhomboidea* is native to Africa, Asia, and Taiwan and was first collected on O'ahu in 1895. *T. semitriloba*

Look for: A small, burr-bearing shrub with hairy, irregular, chewed-up leaves in upper dry and mesic forest.

comes from Mexico, South America, and the West Indies and was observed on Maui in 1910. Both species are now found on most of the large Islands in dry and lower mesic forest.

CASTOR BEAN
Ricinus communis
Euphorbiaceae (crown of thorns, poinsettia, cassava, coral plant)

These weedy aliens are large shrubs or small trees, 10 to 12 feet tall or more, with large leaves about a foot across. The leaves have 6 to 11 main veins radiating from the stem, each acting as the midrib of a sharply pointed, toothed lobe. The stem is attached inside the margin of the back of the leaf, rather than at one edge. Leaf stems and the branches of the plant are hollow. The fruit occur in clusters and are covered with prickly spines. They are about ¾ of an inch in diameter. The seeds of the plant are violently poisonous. The ricin obtained from them is one of the deadliest toxins known, and has been considered to be an agent that might be used in a terrorist attack. The plants are probably native to Africa and were introduced into Hawai'i before 1819, perhaps as garden ornamentals or for the useful oil that is obtained from the seeds. They are now common in low elevation dry areas on all the major Islands.

JOHN HOOVER

Look for: Large thick-stemmed shrubs with large leaves with many pointed lobes arranged like the (unequal) spokes of a pinwheel, and clusters of round spiny fruit.

INDIGO
Indigofera suffruticosa
Fabaceae (golden shower tree, kiawe, lentils, beans)

These plants are erect, open, rather scraggly shrubs 3 feet tall or more with opposite odd-pinnate compound leaves. That is, 9 to 17 paired oval leaflets, about ½ to 1 ½ inches long by ⅛ to ½ inches wide, are arranged along a stem with the odd leaflet at the end. Spikes of small salmon colored pea-like flowers give rise to clusters of small curved bean pods, each about ¾ inch long, that emerge from a common point, often resembling tiny hands of bananas. This species probably came from the American tropics originally, but was introduced to Hawai'i very early, having been first reported here in 1779. A Dr. Serriere is said to have brought indigo from Java to grow in Hawai'i in 1836 and was able to produce a good quality of dye. The industry did not persist, but the plant now grows wild in dry disturbed areas.

KEN SUZUKI

Look for: An open, rather scraggly shrub with compound leaves with an odd number of leaflets and clusters of small, curved bean pods looking like tiny hands of bananas.

Indigo is the material that was used to dye blue jeans and other blue fabrics before the development of synthetic dyes. Several species of this plant are known around the world, and the ancient Egyptians knew and used it. Indigo was a major plantation crop in the southern colonies before the Revolutionary War although its importance in the economy waned as cotton became dominant. To obtain the dye, plants are fermented in water and then the solution is drained off and agitated for several hours. The dye then settles to the bottom as a bluish mud.

JAMAICA VERVAIN
Stachytarpheta spp.
Verbenaceae (lantana, fiddlewood, verbena, vitex)

There are 4 very similar species of this weedy alien plant in Hawai'i. They come from tropical and subtropical America, and the earliest arrival was probably cultivated here before 1871. They are low, branched, sprawling herbs or small shrubs of the dry to mesic forest. The leaves are opposite, toothed and with a rough, corrugated surface. The most striking feature of these common weeds is the long spindly flower stalk which often extends for a foot above the top of the plant and bears a few small but attractive blue flowers,

Look for: A coarse-looking weed with a long flower stalk bearing a few lovely blue flowers part way along its length, that have a flavor like mushrooms.

JOHN HOOVER

about ½ inch in diameter, that are usually part way down the spike. The flowers taste somewhat like mushrooms.

KLU

Acacia farnesiana
Fabaceae (beans, peas, monkey pod, indigo).

The klu bush is a scraggy, viciously thorny shrub of very dry areas along our leeward coastal hills. The young branches have a zigzag form, and unlike the koa, which is also an acacia, it has true leaves which are alternate and pinnately compound, with stemlets arranged in pairs along the main leaf stem, and small oval leaflets in pairs along these. The bush bears yellow, very fragrant, puff ball flowers, followed by a lumpy form of bean pod. Young twigs of klu are

Look for: Zigzag thorny branches, compound acacia-type leaves, and yellow, fragrant puff ball flowers.

a reddish brown, unlike the green twigs of kiawe, which grows in the same area and is similar when young, but which generally grows to be a tree and has greenish-yellow flowers in cylindrical spikes.

The plant is a native of tropical America. Klu was probably introduced to Hawai'i well before 1860, possibly to serve as the basis for a perfume industry. Such an industry, producing cassie perfume, operated in the south of France, but it was not a success here, and we are left with this weedy shrub as a relic of the attempt.

KOLOMONA, SCRAMBLED EGG TREE
Senna gaudichaudii
Fabaceae (beans, monkey pod, kudzu, royal poinsiana)

There are 8 members of this genus in Hawai'i, 7 naturalized introductions and the one above which is indigenous, also being found on some of the islands of the South Pacific. Several of the species are called "kolomona". This plant is a shrub, 2 to 16 feet in height with even-pinnate compound leaves in which 4 or 5 pairs of leaflets are arranged along the stem. These are 1 to 3 inches long and about ¾ inch broad and usually covered with fine hairs. Clusters of greenish white to pale yellow flowers are followed by drooping flat pods about 5 inches long and ½ inch wide containing reddish brown seeds in narrow compartments in a single row. The shrub is found in arid lowlands and along the coast on all of the larger islands.

Look for: A dryland shrub with even-pinnate compound leaves, the leaflets being much larger than in koa haole, and clusters of greenish white to pale yellow flowers, followed by typical flat bean pods.

Another member of this genus that may attract your attention is *S. surattensis*, or scrambled egg tree, which is also called kolomona. This is a small tree up to 20 feet tall with similar compound leaves as much as 7 inches long with 6 to 10 paired leaflets. The oval leaflets are 1 or 2 inches long and an inch or less wide. Clusters of bright yellow flowers, each about an inch in diameter, give the plant its common name. These are followed by flat pods about 5 inches long containing a row of small seeds in narrow compartments. This plant was reported in the wild in Hawai'i before 1871. It is not certain where it originally came from—possibly Australia, India, or southeast Asia. It has now made itself at home in low elevation disturbed sites around the major Islands.

KEN SUZUKI

Look for: A small, spreading tree with compound leaves and clusters of yellow blossoms that look like helpings of scrambled eggs.

KULUʻĪ
Nototrichium sandwicense
Amaranthaceae (pāpala, cockscomb, amaranth)

There are 2 endemic shrubs in this genus in Hawai'i. The one above is the most widespread, though still rare. It is a shrub or occasionally a small tree with opposite, pointed oval, silvery gray-green leaves about 1 to 4 inches long by 1 or 2 inches wide that are covered with white hairs, especially on the lower side. The fruiting body is a catkin-like cluster of tiny scaly flowers in stout, drooping spikes that are also hairy. The plant is found in scattered locations in open dry ridges, lowlands and lava fields on all the major islands. It has recently found some use as a landscape plant in xeriscape (dry land) gardens. The other species, *N. humile*, is similar but has a more slender flower spike. It is found in a few locations in the Wai'anae Mountains on northwestern O'ahu.

Look for: A shrub of the dry lowlands with pointed oval silvery leaves and compact scaly catkins all covered with white hairs.

LANTANA
Lantana camara
Verbenaceae (vervain, vitex, fiddlewood, verbena)

JOHN HOOVER

This is one of the most obnoxious weedy shrubs that the hiker is likely to encounter. It can form dense, sprawling intertwined masses of branches up to 10 feet tall that crowd out all other vegetation and appear to have an allelopathic effect—producing some chemical that inhibits the growth of neighboring plants. In addition, the twigs and branches bear short but nasty curved thorns that cling to skin and clothes and dig in deeper if you try to pull away. Lantana also has a pungent odor. The coarse rough opposite leaves of this plant are pointed ovals 1 to 5 inches long by 1 or 2 wide with blunt teeth along their edges and prominent veins. The small 4-lobed flowers occur in hemispherical clusters and are yellow when they first open in the center of the cluster, turn pinkish and then deep lavender as they age, so that these colors form concentric rings around the yellow centers, producing a striking and attractive effect. White colored varieties also occur. Flowers are followed by small black berries that are spread by birds. The

Look for: A scraggly, odoriferous shrub with course, rough leaves, nasty recurved thorns, and lovely round clusters of multicolored flowers, yellow in the center and becoming lavender toward the edges.

45

plant is probably native to the West Indies but is now found world wide in the tropics. It was introduced into Hawai'i in 1858, probably as a garden ornamental and has since spread to dry, mesic, and wet habitats on all the Islands.

Lantana was such a pest in pastures and on trails that 23 different insects that feed on the plant were introduced in 1902 to discourage its growth. Thus many plants that you will see will have galls on the stems or show evidence of leaf miners or other damage. These insects have had some impact on the population, but there is still a lot of the iniquitous lantana around, much of it all too healthy!

MICKEY MOUSE PLANT

Ochna thomasiana
Ochnaceae (I don't know of any familiar relatives)

These plants are shrubs up to 15 feet tall. The leaves have little or no stem and are closely attached to the twig. They are 2 or 3 inches long, pointed ovals, with fine teeth along the edges. Clusters of short-lived 5-petalled yellow flowers about an inch across are produced on short lateral branches. The most distinctive feature of the plant is the fruit which consists of several oval glossy

KEN SUZUKI

Look for: A shrub with finely toothed leaves attached directly to the twig and red, flower-like structures bearing glossy black oval seeds, in dry and mesic forest.

black seeds, about ¼ inch long, that are borne on a bright red flower-like receptacle. This fruit sometimes resembles the head and large round ears of the cartoon mouse from which the plant gets its common name. The oily, protein-rich seeds are eaten by introduced birds which spread them, so that seedlings are a major nuisance in the garden and the plant is becoming an increasingly serious pest in dry to mesic forest on O'ahu and presumably the other Islands where the plant is cultivated. The shrub is a native of east tropical Africa and is widely cultivated as an ornamental. It was first reported in the wild in Hawai'i in 1998.

PĀNINI, PRICKLY PEAR

Opuntia ficus-indica
Cactaceae (cacti)

This introduced plant with its large oval succulent spiny plate-like segments will be familiar to almost everyone. The cactus bears yellow or orange flowers, about 3 inches in diameter, followed by 3 inch long yellow to purple fruit. It is found in the wild as small plants, shrubs, and even small trees in dry areas on all the Islands. It is probably native to Mexico and is thought to have been introduced to Hawai'i by Don Francisco de Paula Marin from Acapulco before 1809. Rows of this cactus may have been planted to form a barrier to contain domestic stock (pā-nini means "fence wall"), but it has shown a tendency to spread and form extensive dense clumps on ranch land, eliminating a large amount of pasture. To control this plant, a moth, *Cactoblastis cactorum*, whose larvae feed on the cactus was introduced into Hawai'i as a biological control agent and seems to have been successful in greatly reducing the population of pānini, so that it no longer appears to be a serious pest.

KEN SUZUKI

Look for: A typical prickly pear type of cactus, with large oval spiny plate-shaped segments.

The ripe fruit, after the spines are removed, can be eaten raw, dried, cooked into a paste, made into candy or syrup, or fermented to make an alcoholic drink.

PŪKIAWE
Styphelia tamiameiae
Epacridaceae (I don't know of any familiar relatives)

This is an indigenous plant which is also found in the Marquesas Islands. It is a small to large shrub with very numerous, very small, oval needle-like leaves that spiral around the stem. They are gray green above and lighter underneath. The flowers are minute and inconspicuous, but the clusters of small round fruit, each about ¼ inch in diameter, are quite attractive, ranging in color from white through shades of pink to maroon, and sometimes mottled with more than one color. Pūkiawe

Thomas H. Rau

Look for: A very common dryland shrub with very small, pointed oval leaves and clusters of small round white to maroon berries.

47

is a common shrub of dry and mesic forest and ranges upward even into bogs and alpine shrublands. This is a very hardy, highly adaptable species.

Pūkiawe can be confused with several introduced weedy species called manuka or Australian or New Zealand tea trees. The foliage of these plants is somewhat similar, but the invasive weed species have numerous and conspicuous 5 petalled white or pink flowers about ⅝ inches across and dry capsules instead of colorful fruit. Pūkiawe berries were eaten by the native nene, a small goose that now lives in the dry uplands of our higher islands. The Hawaiians used the wood of the shrub for kapa anvils, and its leaves and fruit in lei. Their most interesting use for the plant however, stemmed from their belief that it could modify mana, the spiritual force or aura of a person. When they executed a criminal, they wished to ensure that the ghost would not return to seek revenge, and so they would cremate the person on a fire of pūkiawe in the belief that this would drive out the mana from the bones, where it was thought to reside, and render the ghost harmless. Hawaiian culture was the most rigidly stratified of any in Polynesia, and it was believed that the highest chiefs and chiefesses were people of great sanctity. These, as among the royalty of the ancient Incas and Egyptians, were the offspring of full brother-sister matings, which avoided contamination by the blood of lesser lineages. Such people were obliged to remain in their own compounds during daylight hours, since a shadow cast by one of them on a piece of ground would render that plot forever sacred, and unavailable for any mundane use. Even the lesser aliʻi (nobles), who carried out the practical chores of governing, were thought to have powerful mana which would be dangerous to bring into contact with that of mere commoners and could cause injury to both sides. Thus when it was necessary in the course of business for one of the aliʻi to mingle with the lesser breeds, he (or presumably she?) would enter a special house and bathe in the smoke of a pūkiawe fire while the proper chants were recited and a dispensation invoked to reduce the mana and make contact with the lower orders possible. Further ceremonies restored the mana later, after the business had been completed.

SOURBUSH, PLUCHEA
Pluchea symphytifolia
Asteraceae (ragweed, chamomile, sunflower)

KEN SUZUKI

Most people just call this shrub "pluchea". Only botanists are likely to mention it at all. It is an erect many branched odoriferous shrub, up to 14 feet tall, with alternate wooly oblong leaves, 2 to 8 inches long by 1 to 3 inches wide. The leaves are gray green above and lighter in color beneath. The small

drab whitish flowers occur in large. flat-topped clusters. These develop into tiny fluffy wind-borne seeds. The shrub is native to Mexico, the West Indies, and adjacent coasts. It was first noticed in Hawaiʻi about 1931 and has become an undesirable widespread weed in dry areas and extending up into mesic and wet forest.

ʻŪLEI

Osteomeles anthyllidifolia
Roseaceae (roses, strawberries, raspberries)

ʻŪlei is an indigenous plant that is very common in the dry areas where many of our hikes begin. It fades away when trees that can shade it begin to appear, although it may occur again in subalpine shrublands. ʻŪlei is also native to the Cook Islands and to Tonga. The plant is a sprawling, viny shrub with very distinctive leaves. We say that these leaves are odd-pinnately compound, which means that pairs of small leaflets occur on opposite sides of the leaf stalk, with the odd one at the tip. The small white flowers look somewhat like wild roses, and the white fruit is shaped much like a miniature rose hip. The branches pile up to form dense tangles along the side of the trail. The scientific name rolls melodiously off the tongue, once you are accustomed to it.

The wood of the ʻūlei is very hard, and the Hawaiians used it for spears and for the ʻōʻō, the digging stick that was their primary tool in agriculture. The

flexible younger branches were made into the hoops for hand nets in fishing and for fish traps and baskets. A lavender dye was extracted from the berries, and, according to one source, these, like rose hips, were eaten.

Look for: Dense, sprawling mounds of a viny shrub with odd-pinnately compound leaves. White flowers and miniature white rose hips are present in season.

TREES

ALAHE'E
Psydrax odorata
Rubiaceae (gardenia, coffee, cinchona - quinine, many natives)

These indigenous trees are also found on the islands of Micronesia and the south Pacific. They are small trees up to 20 feet tall with smooth white bark and horizontal branches bearing glossy dark green opposite leaves, somewhat resembling those of mock orange. The leaves are pointed ovals 2 to 3 inches long by 1 inch wide. Clusters of small white fragrant flowers appear on the branch tips in season and are followed by round black berries about ⅓ inch in diameter. Alahe'e is common in dry and mesic forests in Hawai'i.

These trees are being used in landscaping in towns around the Islands. The Hawaiians found the hard, durable wood suitable for digging sticks or 'ō'ō, as well as for spears, hooks for shark fishing, and adze blades that could be used to cut softer wood, such as that of kukui or wiliwili. A black dye was prepared from the leaves of the plant. "Ala" means fragrance, and "he'e" is octopus or to flow, so the name refers to the way the fragrance of the flowers slithers down the breeze.

KEN SUZUKI

Look for: Small erect attractive white-barked trees with paired glossy dark green leaves and fragrant clusters of white flowers or black berries, in season.

CHRISTMAS BERRY, BRAZILIAN PEPPER
Schinus terebinthifolius
Anacardiaceae (mango, poison ivy, sumac, cashew, pistachio)

Christmas berry was introduced from Brazil as an ornamental sometime before 1941. It is now one of the worst of our invasive weed trees in dry forest areas, forming dense tangles of low branches under which almost nothing else will grow. It occurs as a shrub in very dry areas and extends well up into the mesic forest as a tree. The tree is a low growing, gnarled, spreading plant with gray bark that becomes rough and furrowed with age. Dead limbs accumulate under its thickets creating a serious fire hazard. It gets its name from the large clusters of attractive small red berries that appear late in the year and are often used in Hawai'i in wreaths or other Christmas decorations. These are poisonous, however, and should not be eaten. The leaves alternate along opposite sides of the twig and are odd-pinnately compound—that is, 2,3, 4 or more pairs of ¾ inch to 3 inch long oval leaflets are arranged on opposite sides of the leaf stem, with the odd one at the tip. This leaf form, with such large leaflets, is quite distinctive among Hawaiian plants, and the identification can be confirmed by crushing one of the leaflets and smelling the pungent, peppery fragrance, which accounts for the alternate name for this plant. Members of this family are known for their ability to cause contact allergies, and some people may become sensitive to this species.

Thomas H. Rau

Look for: A low, gnarled, spreading dry forest tree or shrub, with rough gray bark and compound leaves that have a peppery odor when crushed.

COMMON GUAVA
Psidium guajava
Myrtaceae ('ōhi'a, mountain apple, eucalyptus)

This small tree is native to the American tropics and was probably introduced into Hawai'i very shortly after its discovery by Europeans, in the early 1800s, by Don Francisco de Paula Marin. It is now established in disturbed habitats in dry and mesic forests throughout the Islands. Like the strawberry guava, the tree has attractive smooth reddish-brown bark. Its hairy twigs are often square in cross section and they bear opposite leaves that are oval with blunt points, 2 to 6 inches long by 1 to 3 wide. The leaves are not smooth and glossy like those of strawberry guava, but are rough and a dull matte green with prominent veins, that are particularly obvious on the back of the leaf. The fruit is about the size and color of a lemon, but soft, and usually pink (sometimes white) inside. It is quite edible and is used to make juice, jams, jellies, and

Look for: A tree with smooth, reddish-brown bark, coarse, dull green leaves with prominent veins on the back, square twigs, and fruit the size and color of a lemon.

other consumables. Birds, pigs, and other animals also enjoy it and help to spread the seeds. The fruit provides a breeding ground for several species of alien fruit flies that are serious agricultural pests.

This tree can be a serious pest in the wilds and often invades grazing lands. Although it is not a native, it has been in the Islands long enough that the Hawaiians made some medicinal use of it. The leaf buds were chewed by a mother with a young infant and the pulp given to the child to treat diarrhea. Tap roots of young trees and bark from older ones formed part of a mixture used to treat dysentery. The leaf buds also formed a component in a concoction applied to sprains or cuts.

FORMOSAN KOA
Acacia confusa
Fabaceae (monkey pod, shower trees, lima beans, alfalfa).

Formosan koa is an attractive small tree with light colored bark and slightly curved, sickle-shaped leaves. It comes originally from Taiwan and the Philippines and has been extensively planted in dry to mesic areas of Hawai'i. The tree was brought to Hawai'i about 1915 by the Board of Agriculture and Forestry, and also by the Hawai'i Sugar Planters Association. The "leaves" are not true leaves, but like the "leaves" of our mature native koa are modified leaf stems. These are called "phyllodes" by botanists. If you look closely at the "leaf", you will see that there is no central midrib with side veins branching from it, but rather a number of parallel veins extending from the base to the tip of the leaf. The flowers are small round yellow puff balls about ½ inch across and these are followed by typical bean pods. Hawaiian koa has paler, cream-colored flowers. Unlike our native koa, Formosan koa never displays true, acacia-like leaves. Its leaves are usually smaller and less curved than those of the native koa, being about 2 to 4 ½ inches long compared to 3 to 10 inches for the native. Another major difference is that Formosan koa leaves often stick straight out of the twig in all directions, so that their tips roughly describe a cylinder and many leaves

KEN SUZUKI

Look for: A small tree with light colored bark and alternate sickle-shaped leaves with no midrib and bright yellow puff ball flowers.

are erect, not hanging down. In the native koa, on the other hand, most leaves (phyllodes) tend to hang down, and the leaf blade lies in a vertical plane so that it exposes the minimum surface to the mid-day sun. While there is much overlap, Formosan koa generally grows in drier areas than the native koa.

IRONWOOD, CASUARINA

Casuarina spp.
Casuarinaceae (There are about 50 species in this family, all in the genus *Casuarina*. No other close relatives.)

Ironwood is one name given to 2 similar species of trees, native to Australia and a few of the Pacific islands near it, that have been extensively planted in Hawai'i. The tree can grow to be quite large. These trees are very tolerant of salt spray and have been used widely for beach plantings, but have also been planted on ridges in dry areas of the mountains. Casuarina were probably introduced to Hawai'i about 1882. They are sometimes referred to as "pines" because of the long, needle-like twigs that assume the functions of leaves, and even have what look like small cones. However, in botanical terms, they are actually dicotyledonous angiosperms, meaning that they are more closely related to walnuts or string beans

Look for: A tree with long, drooping, needle-like "leaves" and small cones.

than to pines. If you examine the "needle" carefully, you will see that it does not occur in bundles like pine needles but is attached individually directly to the twig, and that it is divided into segments. Pull a needle apart at one of these joints. Notice the tiny triangular bristles on one side of the break. These are all that remain of the true leaves of this plant. Ironwood groves tend to expand slowly as suckers grow up from the roots and form new trees. The shed needles form a soft carpet, like pine needles, and are pleasant to walk or rest on, but they seem to have allelopathic properties—they poison other plants and very little other vegetation is found under a stand of ironwoods. For this reason they can be a serious problem in our forests where they crowd out native plants.

Ironwoods form a nitrogen-fixing association with bacteria in the genus *Frankia*. This may help them to thrive in the nutrient-poor conditions often found in tropical soils. Although ironwoods were not found in Hawai'i when the first Polynesian settlers arrived, they were apparently familiar with the tree since it grows on some of the islands further south. In these islands, Norman Scofield tells me, they called it "toa" meaning "brave warrior", since it was such a large, sturdy tree. On settling in Hawai'i, they gave the same name to the largest, strongest tree they encountered here, and with the passage of time, this became "koa".

JAVA PLUM
Syzygium cumini
Myrtaceae ('ohi'a, guava, eucalyptus, mountain apple)

Java plum is a medium sized tree with light gray bark that is very common in dry to mesic valleys and disturbed forests. The bark is smooth in young trees but may become furrowed with age. It is a native of south and southeast Asia and was probably introduced to Hawai'i for its fruit late in the 19th century. The leaves of this tree are simple, lance-shaped, opposite leaves about 3 to 7 inches long. Young leaves are often reddish in color. The flowers are found in clusters of inconspicuous white blooms

Look for: A medium sized, gray barked tree with opposite lance-shaped leaves that is very common in dry to mesic forests.

that give rise to abundant purple-black olive-sized fruit, which often litter the ground under the trees with purple blotches in season. The fruit is edible but often tart.

KIAWE
Prosopis pallida
Fabaceae (koa, pea, alfalfa)

This common lowland tree is a species of what is called "mesquite" in the southwestern U. S. Reportedly, all the trees in Hawai'i descend from a single seed from the Royal Gardens in Paris that was planted on the grounds of the Catholic Mission on Fort Street in Honolulu in 1828 by Father Bachelot. The trees are native to northwestern South America. Kiawe can grow to 60 feet or more and is armed with long, sharp thorns that readily puncture bicycle tires or the soles of zoris, as I can attest from personal experience. It often grows as a gnarled, twisted tree with prominent, contorted ridges on the trunk. Leaves and thorns emerge from a common node on the twig and the greenish twig changes direction slightly at each node, giving it a zigzag form like that of the related klu bush. The leaves are doubly compound, with 3 or 4 pairs of stemlets branching from the leaf stem, each bearing 6 to 15 pairs of small leaflets about a ¼ inch long by ⅛ inch wide. Tiny pale yellow-green flowers are borne in 3 to 5-inch long cylindrical spikes, and these are followed by yellow-brown pods 3 to 10 inches long. The small brown seeds are contained in a sweet gummy pulp and the pods are an excellent feed for cattle. My wife recalls that when she was young, the children would collect the pods to sell to cattlemen, receiving a few cents for a large bag full. The cattle relished the beans and produced rich milk when feeding on them. The tree is now abundant in low dryland areas in all the Islands.

Look for: A common lowland tree with a gnarled, sinewy trunk and branches, sharp thorns and delicate compound leaves.

Kiawe is an excellent honey plant, and at one time sugar and honey were the major export products of Hawai'i, with 200 tons of the latter being shipped every year. A great deal of very good charcoal was also produced from the wood. Young kiawe may resemble klu, a shrub of the same habitat, but klu twigs are reddish brown, not greenish, its flowers are bright yellow puffballs, and its bean pods are short, lumpy, and dark brown.

KOA HAOLE
Leucaena leucocephala
Fabaceae (indigo, wiliwili, peas, and all the great bean family)

Koa haole, or "foreign koa", has leaves that look like the true leaves of native koa seedlings before the latter are replaced by the sickle-shaped phyllodes. The small, ½ inch long leaflets of koa haole are longer and more pointed than those of koa, however. This alien tree comes from Central America and the Caribbean, but was introduced to Hawai'i from Guam or the Philippines sometime before 1830. These are very drought-tolerant small trees, up to about 30 feet in height, that were introduced for reforestation and cattle browse. They form dense thickets in the driest parts of the dry forest, clear down to sea level. Many pairs of small leaflets are arranged along stemlets that lie opposite each other along the main leaf stem. These (doubly) compound leaves alternate with each other on opposite sides of the twig. Small white puff-ball flowers emerge from the junction of leaf with stem, and produce clumps of long, flat bean pods with many seeds. The plants have no thorns.

Cattle can feed on the leaves of koa haole, but they contain an amino acid, mimosine, which is not found in typical proteins, and which is toxic to horses and pigs, causing them to loose their hair and hooves. Cattle will also become ill and lose hair if fed too much of the plant. Although koa haole is an aggressive weed of dry forests in Hawai'i, crowding out virtually all other plants in areas where it flourishes, some of the 22 species of *Leucaena* which are native to the Americas from Peru to Texas are of considerable value to the people in these lands. In some areas the young pods are eaten for their garlic flavor. They also fix nitrogen and enrich the soil, and are often used as shade trees for shade-grown coffee. The trees can be valuable as firewood, chip wood, and green manure. The *Leucaena* are among the world's fastest growing trees, and Dr. James Brewbaker and his colleagues at the University of Hawai'i have done a great deal of work to select favorable varieties and hybrids to enhance their value for the above purposes. One of their hybrids will grow to be a tree more than a foot in diameter in 12 years! Firewood from such trees can help spare native forests and provide a carbon-neutral source of fuel.

JOHN HOOVER

Look for: Small, dry forest trees without thorns and with acacia-like compound leaves. When flowering, there will be white puff-ball flowers about ¾ inch in diameter followed by long, flat bean pods.

LAMA
Diospyros sandwicensis and *D. hillebrandii*
Ebenaceae (ebony, persimmon)

There are two endemic species of lama in Hawai'i and both are worth noticing. The first, *D. sandwicensis* is one of the commonest native trees in the dry forest. It appears in very dry areas and extends up to the wet forest. This tree has dark brown to black bark which is not furrowed or flaky, but is rough to the touch, and drab dark gray-green pointed oval leaves about 1 to 3 inches long that alternate on opposite sides of the stem. The foliage is often quite dense. Young leaves have a pinkish tinge. The fruit is ½ to ¾ inch long and turns orange to red when ripe. When fully ripe, it is sweet but with an astringent tang vaguely reminiscent of its relative, the persimmon. It is best not to eat any strange fruit, of course, unless you are absolutely sure of what it is! Lama is often infested with a species of tiny mite (an Eriophyid) that produces clusters of finely branched growths 2 inches or more in diameter on the twigs or branches.

"Lama" means "light" or "lamp" in Hawaiian, and, as in English, had a figurative meaning of "enlightenment". The thick sapwood of lama is white and was used in religious contexts. Sacred areas were fenced with lama wood (Pālama— an enclosure of lama). In the hālau hula, the structure consecrated to the forest goddess Laka (goddess of hula and wife and sister of Lono, one of the 4 major Hawaiian deities) and reserved for the use of hula dancers and trainees, a block of lama wood was wrapped in a piece of yellow kapa scented with 'ōlena (turmeric) and placed on the alter. This represented Laka. Lama or 'ōhi'a wood was generally used for the construction of the buildings forming part of a heiau (temple).

The other lama species, *Diospyros hillebrandii*, is much less common and occurs in less dry areas, but I am always pleased when I see it. It has larger leaves, and the young leaves are often a beautiful shade of pink. The older leaves repay a close look, and to me are one of the gems of our native forest. Imagine that a skilled jade carver has taken a piece of dark green jade and carved it into the form of a leaf. On the surface there is an intricate network of veins in raised relief polished to a high gloss, in contrast to the matt green of the leaf blade between. Such is one of nature's hidden jewels. This is the leaf of the "other" lama. Look closely, think small!

Look for: A tree with black bark and alternate gray-green leaves.

JOHN HOOVER

LOGWOOD
Haematoxylum campechianum
Fabaceae (clover, indigo, mesquite, and all the legumes)

This is a thorny tree that was introduced from tropical America and planted sparingly in low dry sites on O'ahu and Hawai'i. It was probably brought in as an ornamental sometime before 1871. The trees have compound leaves with an even number of leaflets in 3 to 5 pairs along the leaf stem. Each leaflet is an inch or less long, oval, with a rounded or dimpled tip, broader near the end than at the base. Logwood can most readily be seen along the Kuliʻouʻou trail on Oʻahu.

This was once a very valuable tree since the heartwood was the source of a red dye, hematoxylin, that was used for stains (in medical tissue studies, among others) and in inks, before synthetic dyes replaced it. The Scots pirates that founded British Honduras (now Belize) turned to the harvest of mahogany and logwood when piracy became

Look for: Rough, thorny trees with even-pinnate compound leaves that have leaflets that are broader toward the tip and often a flat or even indented tip.

too risky, and their descendants from matings with the African slaves they brought in to do the work, make up the largest component in the ethnic mix in Belize today. The heavy, reddish brown wood can also be made into fine furniture. Now the tree is primarily planted as an ornamental.

LONOMEA
Sapindus oahuensis
Sapindaceae (ʻaʻaliʻi, lychee, longan, ferntree)

Lonomea is a tree up to 50 feet tall with alternate lance-shaped leaves with prominent yellow midribs. They are 3 to 8 inches long. The tree is rarely seen in flower or fruit, but if present, the fruit is an oval black berry about an inch long that usually has a tiny aborted twin attached to the stem end. This tree is very similar to the native olopua and grows with it in similar habitat on Oʻahu and Kauaʻi, but lonomea has alternate leaves while those of olopua are opposite. Lonomea was the name used for this tree on Kauaʻi. It was also called āulu or kaulu, but since these names were used for another native

KEN SUZUKI

58

tree as well, I prefer to call it "lonomea". This tree is endemic to Kaua'i and O'ahu were it is found in dry to mesic forest.

An indigenous relative, *S. saponaria*, occurs on the Big Island as well as in Mexico, South America, Africa, and many Pacific islands. This tree, like all others in the genus except lonomea, has even-pinnately compound leaves. It has 3 to 6 pairs of oblong leaflets 3 to 6 inches long, that taper to sharp points. This tree is also called "soap berry" as the pulp of the fruit yields a lather good for washing hair or fine fabrics.

Lonomea seeds were strung into lei, and had some medicinal use as a cathartic, although they should be used with caution as they may be poisonous. The wood was used in house construction and for spears.

Look for: A tree with alternate lance-shaped leaves with prominent yellow midribs in dry and mesic forest on O'ahu and Kaua'i.

OLOPUA
Nestegis sandwicensis
Oleaceae (olive, ash, jasmine, pīkake)

This is the only native plant in its family in Hawai'i. It is an endemic shrub or tree with opposite lance-shaped leaves about 3 to 6 inches long with yellow stems and midribs. I have seen it in flower or fruit only occasionally. The flowers are small and inconspicuous, yellow-green in color, and cluster at the base of the leaves. They are followed by oval green berries that turn black as they mature. The tree is scattered in dry and mesic forest on all the larger Islands. Olopua resembles another endemic tree, lonomea, which grows in similar habitat on Kaua'i and O'ahu. The trees can readily be told apart however, since olopua leaves are opposite ("o" for olopua, "o" for opposite) while those of lonomea are alternate.

The hard durable wood of olopua was valued by the Hawaiians for adze handles and other tools. It was used in rasps to shape

Look for: A tree of dry and mesic forest with opposite, lance-shaped leaves that have yellow stems and midribs.

fish hooks and for house posts and poles. It was also favored as firewood, since it burns with a hot flame, even while still green. The tree was a common host to our native land snails. Olopua is often seen as more of a shrub than a tree, with many small branches sprouting from the base. This is not its original form, but is due to the attack of the black twig borer, *Xylosandrus compactus,* an insect originally from Japan that was accidently introduced into Hawai'i in about 1931. It injures many of our native trees by boring into the twigs and infecting them with an ambrosia fungus on which its larvae feed. The fungus often causes a tree to decline or die. When this borer kills many of the twigs at the top of the olopua, the tree responds by sending up numerous suckers from its base, which gives it the bushy appearance. The borer attacks coffee plants and other agriculturally important trees and shrubs, but is particularly damaging to many of our natives. Koa suffers from it, and it has led to the virtual extinction of what was once the most massive of our native trees, the mēhamehame or *Flueggea neowawraea*, a tree that once grew to be over 100 feet tall and 7 feet in diameter.

STRAWBERRY GUAVA
Psidium cattleianum
Myrtaceae (eucalyptus, cloves, allspice, mountain apple, 'ōhi'a)

Strawberry guava is a lovely tree with smooth warm light brown bark, glossy green leaves, and delicious red fruit. It is also the worst of the pernicious invasive weed trees that are crowding out our native plants, although if *Miconia* can not be contained in its current strongholds on Maui and the Big Island, it may loose that dubious honor. You may hear it called "waiawī" also, although usually this name is restricted to one of the less common yellow-fruited varieties. Strawberry guava was probably introduced into Hawai'i from coastal Brazil in 1825 on the *HMS Blonde*, a ship that brought a variety of seeds and fruit and nut trees from England and Brazil to the Islands. The tree has since spread throughout the Islands. It is found as a shrub in very

KEN SUZUKI

dry areas, flourishes in somewhat damper areas in the mesic or intermediate zone, and penetrates into the wet forest. The bark has the color of coffee with cream, with grayer blotches where a strip has pealed off, and is smooth, making a comfortable grip for the hands when used as a walking stick. The wood is tough enough to make it a good choice for this purpose. Do not cut or damage any native tree for this or any other purpose, though. They have a tough enough time coping with invasive weed trees, introduced insects, and other novel hazards of life as it is. The fruit of the strawberry guava is about the size of a quarter and is usually red, although yellow varieties are not uncommon. The ripe fruit is much better eating than that of its close relative, the common guava, in my opinion, and has a vaguely strawberry-like flavor. Once seen, the bark is quite distinctive and can be mistaken for no other local plant, except for the common guava, (*P. guajava*) which is often found in the dry forest also. The common guava has a lemon-sized yellow fruit, and coarse, non-glossy, often hairy leaves with prominent veins branching from the midrib, unlike the smooth, glossy leaves of strawberry guava.

A word on invasive plants: I am sometimes told that a "weed is just a plant out of place", implying that whether a plant is a weed or not is just a matter of human preference. This may be true of the weed in a garden, but egregiously understates the menace of invasive plants in Hawai'i.

A plant in its native habitat is assailed by a host of herbivores and parasites—stem borers, leaf miners, sap suckers, gall formers, leaf chompers, bark borers, and root feeders, just to mention the insects. And viruses, bacteria, fungi, nematodes, and mites drain the plant's resources also. A large fraction of the energy that it captures from the sun must go into systems to resist the attacks of these creatures and repair the damage that they do. In addition, it is advantageous for most tropical plants to grow widely scattered over the landscape to make it more difficult for the pests to find a new host after reproducing on the old one. When a seed from a plant is taken to a distant land and planted, however, it generally escapes most of these parasites. If soil and climatic conditions are favorable, it can devote the energy once used in coping with them to growth and reproduction, and thus has an enormous advantage over the local flora, which are still burdened by their pests. Thus the strawberry guava, instead of growing in a bal-

JOHN HOOVER

Look for: A small tree with smooth, light-brown bark and opposite glossy green leaves.

anced, mixed community of plants, has turned large areas of our forests into a monoculture, and you can walk for thousands of yards on many of our trails through monotonous thickets of strawberry guava, unrelieved by any other plant. Thus plants that become invasive are not just plants that people do not want to be where they are (although this is true) but plants that have escaped the natural checks that help integrate them into a balanced biological community, and have run amok at the expense of the community into which they have been introduced. (*Homo sapiens* could be considered a species of this kind, too!) With time, of course, evolution will lead to the development of a host of new parasites, herbivores and competitors, and these invasive species will once more become part of a balanced community. But this may take several hundreds of thousands of years, and I don't have the patience to wait! In the meantime, our fascinating and unique plant communities in Hawai'i are rapidly being replaced by dull monocultures.

A large number of introduced trees have shown a tendency to form groves of a single species—Christmas berry, ironwood, ardisia, rose apple, coffee, hau, eucalyptus, Cook pine, fire tree, koa haole, strawberry guava and bamboo, for example. The ground under these trees is often open with little in the way of shrubs or herbs growing beneath them. Sometimes, as in the case of rose apple, strawberry guava, and bamboo, this may be because they cast such dense shade that most plants can not survive in it. In others, including Cook pines, eucalyptus, ironwood, and, often, strawberry guava, even when plenty of light reaches the ground, there is little other plant life. We think that such plants are *allelopathic*, (roughly, other-injuring) which means they produce compounds, either in their leaf litter or secreted from the roots, that inhibit the growth of other plants, thus discouraging competition.

Strawberry guava spreads easily because seeds from its attractive and abundant fruits are carried long distances by pigs and birds. For awhile, from the mid eighteen hundreds well into the nineteen hundreds, the growth of guavas was controlled in some areas by the extensive harvest of the wood for making charcoal, which was sold for cooking or used in the boilers of the steam engines that carried sugar cane from the fields to the mills. An old charcoal kiln can still be seen along the road in Ho'omaluhia Botanical Garden in Kāne'ohe, O'ahu.

SILK, SILVER, or HE OAK
Grevillea robusta
Proteaceae (proteas, macadamia nuts)

This native of Australia has been used extensively for reforestation in Hawai'i. Only *Eucalyptus robusta* has been planted more frequently. It was introduced in about 1851 and between 1919 and 1959 over 2.2 million trees were planted here. Silk oak will grow in dry areas, but does best in moister, intermediate forest climates. Although it is well established in the wild, at present it does not seem to spread too aggressively or form dense stands that exclude all other plants, and in some dry tracts, does not even replace itself. It has become a pest on some ranch lands on the Big Island, however. This is a medium to fairly large tree with light gray, rough, bark with many closely spaced furrows. The trunk tends to be straight and tall. In season, April or May and extending through the summer, the tree has clusters of striking yellow blossoms. The leaves are a very distinctive feature. They alternate along the stem and are quite large, 6 to 12 inches in length. Dark green above, they are silky and lighter colored beneath. The leaves are compound, with paired leaflets arranged along the central stem, and unlike most compound leaves, the individual leaflets are deeply lobed with sharp points, so that the leaf as a whole has a fern-like appearance.

The silk oak is a good timber tree and has been used for furniture, cabinetmaking, and panels in Hawai'i. It can also be used for paper pulp or veneer. Many

Look for: A tree with an erect, usually straight trunk and large, fern-like leaves. It will have striking yellow flower clusters in season.

trees in the extensive stands in the Honouliuli Reserve on O'ahu have been dying. Studies suggest that the life span of the tree is 60 or 70 years, and since these trees were planted in the 1930s, they are probably dying from old age rather than disease or insect attack.

WILIWILI
Erythrina sandwicensis
Fabaceae (peas, beans, acacia, gorse, indigo)

Wiliwili was once one of the dominant trees of the lowland dry forest, the area now often occupied by the introduced kiawe. This endemic species is medium-sized, with a stout, often crooked trunk clothed in smooth orange-brown bark. It is one of the few natives that may have thorns, frequently having broad conical points, especially on younger trees. The leaves are compound, with three large oval leaflets much like those of string beans. The leaves are commonly shed in fall and winter before the tree flowers. Blossoms are usually orange, but may be red, yellow, white or even a pale green, the

Hawaiian species being unusual in the genus for the variety of colors displayed. The lumpy bean pods shed attractive red or orange beans that were used in lei. Recently, *Erythrina* species in Hawai'i have been attacked by an introduced gall-forming wasp, *Quadrastichus erythrinae*, and many are dying, so the future of our native wiliwili is in doubt.

There are at least 100 species of *Erythrina* and many of them have been used as street or garden trees, in Hawai'i and elsewhere. Ho'omaluhia Botanical Garden in Kāne'ohe, O'ahu has a large collection of them. The wood of the native tree was soft and light in weight, like balsa wood, making it suitable for the floats on outrigger canoes or fishnets. It was also used to make the long variety of surf board, the olo, which could be up to 20 feet long. In historic times, most such boards were made of koa, which made them very heavy to carry, possibly because of the scarcity of wiliwili trees of sufficient size. Charcoal from limbs of the wiliwili was one of the pigments favored for the paint used on canoes.

According to one myth, in which wiliwili were involved, as told by Marie C. Neal (*In Gardens of Hawaii*), there were 4 sisters living in Ka'ū on the Big Island. One was bald, one had scraggy wind-blown hair, and one was hunchbacked, but one, Moholani, was beautiful. After Moholani married, her son was adopted by the gods and raised in heaven. It seems that Moholani's husband was

Look for: A spreading, gnarled tree with smooth orange-brown bark and 3-part compound leaves in dry, lowland areas.

in the habit of talking with the sea sirens, and one day he failed to return from a fishing trip, having gone down into the sea with them. She begged her sisters to help her find him, but they just shrugged off her concern and said, "Good riddance to that big, worthless man!" Her son came down from heaven and in his wrath transformed the sisters into wiliwili trees, the bald one becoming a tree with few leaves, the next a wind-tossed tree, and the last a gnarled and twisted tree. Well into historic times, these 3 trees could still be seen standing by the cave Pūhi'ula at the shore at Pa'ula in Ka'ū. Moholani's husband returned to his wife, and wandered no more, for he feared the anger of his son.

VINES

HUEHUE
Cocculus trilobus
Menispermaceae (I don't know of any familiar relatives)

This slender indigenous vine is common in dry and mesic forest on all the major Islands. It has alternate oval or pointed oval leaves with three fairly marked veins running from base to tip. They are about 2 inches long by an inch wide. The sexes are on separate plants, so you may notice two somewhat different kinds of flowers, but both are small yellowish-white bluntly-tipped star-shaped blossoms. These occur in small clusters at the base of a leaf and are followed by dark blue berries on the female plant, less than ¼ inch in diameter. Huehue is found from the Himalayas through south east Asia and the Pacific islands to Hawaiʻi. It can be confused with young vines of the noxious introduced huehue haole, *Passiflora suberosa*, but the leaves of the latter are much more consistently 3-lobed on older vines, and a thick layer of ribbed corky material develops on mature vines of this weedy plant also.

The Hawaiians probably found this flexible vine useful as a kind of string for a wide variety of purposes. Among other uses, it may have been employed to tie small poles to grass houses and anchor thatch to the structure.

KEN SUZUKI

Look for: A slender vine, often running along the ground, with alternate oval leaves with 3 noticeable veins running from base to tip.

IVY GOURD
Coccinea grandis
Cucurbitaceae (gourd, melon, squash, cucumber)

This introduced vine climbs over other plants by means of tendrils. The alternate leaves are 2 to 4 inches broad and often have a pentagonal shape, although they may also have 3 to 5 more or less deep lobes. The trumpet-shaped flower is 1 ½ to 2 inches wide, white, and with 5 star-like points. This is followed by an oval red pickle-like fruit about 2 to 3 inches long. The vine is native to Africa, Asia, and Australia. It was introduced into Hawai'i from Fiji about 1969 as an ornamental. The seeds are spread by birds and the plant has become a major pest in dry areas on O'ahu and the Big Island, where it clambers over shrubs and trees and smothers them. It is listed as a noxious weed by the Hawai'i Department of Agriculture. The plant resembles our native Sicyos, but the latter have small white blossoms in clusters and small fruit, while ivy gourd bears 1 or 2 larger flowers and a large fruit.

Look for: A vine with alternate pentagonal or lobed leaves and white star-like flowers that smothers other plants, fences, and utility poles in dry areas.

FERNS

DORYOPTERIS
Doryopteris decipiens
Pteridaceae (maidenhair ferns, cliff brake, gold fern)

There are 4 endemic species in this genus in Hawai'i. They are relatively small attractive ferns with fronds up to 1 foot tall. The fronds are deeply lobed and may be as wide as they are long so that the overall form is roughly triangular or sometimes pentagonal. Given their stubby form and intricate lobes, they remind me, with a little imagination, of a snowflake, although without the hexagonal symmetry, of course. When fertile, the spore-bearing structures run along almost the entire margin of the frond. This species is not uncommon in dry shrublands, grasslands, and forest, especially in exposed rocky sites, between 500 and 3000 feet in elevation on all of the main Islands.

KEN SUZUKI

Look for: A small to middling fern with a deeply lobed frond that is roughly triangular or pentagonal in shape in fairly dry, often rocky areas.

GOLD FERN
Pityrogramma (Pentagramma?) austroamericana
Pteridaceae (Cretan brake, maidenhair ferns)

This fern grows in small dense clusters of fronds about 1 or 2 feet tall. The frond is twice divided, pinnately compound, and both the overall shape of the whole frond and of each of the distinctly spaced side branches is like a lance that tapers to a long, sharp point. The most striking feature of the fern, and the one that makes it easy to identify, is the presence of a fine yellow-gold powder that covers the underside of each leaflet. The spore-bearing sori are small black dots that form lines on either side of the midrib of a leaflet, although they are often obscured by the golden powder. These attractive ferns are native to South America and are commonly grown in gardens. It is likely that they were introduced into Hawai'i for this purpose. Gold fern was first collected on Kaua'i in 1903 and is now widespread in exposed dry areas on all the main Islands.

A less common relative, *P. calomelanos*, the silver fern, is also found in Hawai'i, usually in somewhat shadier and moister sites. This fern is similar to gold fern but has a white powder beneath the leaflets. If you slap a frond from

Look for: An attractive fern with doubly compound fronds those leaflets are coated on the underside with a fine yellow-gold powder, of dry exposed habitats.

either of these ferns against a suitable dark surface, a striking copy of the lacy frond structure can be produced. Both of these ferns have been reported to produce allelopathic compounds that inhibit the germination and growth of nearby plants, thus reducing competition for water, nutrients, and light.

MOA
Psilotum nudum
Psilotaceae (no familiar relatives)

This common fern consists of tufts of upright branching green or yellowish stems without obvious leaves. In fact, the leaves have been reduced to tiny arrowhead-shaped scales that cling to the stems. The plant is usually 6 to 18 inches tall. When fertile, numerous 3-lobed yellow balls about ⅛ inch in diameter are scattered along the terminal segments of the plant. The stems are angular, ranging from hexagons to triangles in cross section and divide to form many regular "Y" shaped branches. The fern is indigenous, being found throughout the tropics. It occurs in many habitats, colonizing recent lava flows, rocks, the ground, and the forks of trees. It is often found in greenhouses and gardens, as virtually a weed. The plant is present from fairly dry to moist forest on all the Hawaiian Islands.

JOHN HOOVER

Look for: An erect, branching, leafless plant with angular stems and tiny yellow balls scattered along the terminal segments.

A second species in this genus, *P. complanatum*, is also found in Hawai'i. It is very similar to moa, but has flat, rather than angular stems, and tends to droop. It is less common than moa, but you will sometimes find it growing on trees in the upper mesic and wet forest. Hawaiian children played a game in which Y-shaped terminal sections of moa were interlocked and each child pulled on a pair of the legs until one plant broke. The winner, with the un-broken fern, then crowed like a cock. ("Moa" means "chicken" in Hawaiian). The early Hawaiians used the spores of the plant like talcum powder to pre-vent chaffing beneath a loin cloth. They were also used as a powerful laxative, and, boiled to make a tea, as a treatment for thrush, a yeast infection of the mouth that was apparently common in children. A tea made from the plant was drunk during treatment for asthma or chest pains, and was a component of a mixture used to treat bad breath and a white coating of the tongue.

PĀKAHAKAHA
Lepisorus thunbergianus
Polypodiaceae (laua'e, laua'e haole)

This small fern has simple ribbon-like fronds that taper at both ends. It is about 2 to 10 or more inches in height, and about ¼ to ½ inch wide. The spore-bearing sori are round or oval brown bodies that lie in a row on either side of the midrib of the frond on the upper half of the underside of the blade. This fern is common on trees and rocks in dry, mesic, and wet for-est on all the main Islands. It is an indigenous fern and is also found in China, the Philippines, Japan, Korea, and Taiwan.

Look for: A small fern with a simple unbranched, undivided, ribbon-like blade, tapering at each end, with round spore bodies on the upper half, on rocks or trees in dry, mesic, and wet forest.

MISCELLANEOUS

STARFISH STINKHORN
Aseroe rubra
Basidiomycetes (mushrooms, rusts, smuts)

This is not a very common fungus, but is so striking when you do see it that I wanted to include it. Children find it particularly fascinating. The ground-hugging fruiting body is shaped like a starfish with 5 to 7 double arms forming a star about 3 inches in diameter. It is bright red with a spore-bearing black slime mottling the base of the arms and in a ring around the center. It has a strongly fetid odor that attracts flies which probably help to disperse the spores. Rain may wash off the slime and reduce the odor, however. This fungus is often found in eucalyptus groves in mesic forest in the Fall on all the major Islands. It also occurs widely in the topical Pacific.

Hemmes and Desjardin (*Mushrooms of Hawai'i*) say that the immature bulb is edible before it opens to reveal the red arms.

Look for: You can't mistake it! A ground-hugging red starfish-shaped fungus that really stinks!

Mesic Forest Plants

HERBS

'ALA'ALA WAI NUI
Peperomia spp.
Piperaceae ('awa or kava, black and white pepper, betel pepper)

There are 26 species in this genus in Hawai'i, 23 of them endemic, 2 indigenous, and 1 naturalized. They are usually succulent herbs with virtually invisible flowers and fruit borne on prominent slender erect spikes. Most prefer moist sheltered areas where they grow on the ground or on moss-covered rocks or trees. In many species, the stems

Look for: A low growing plant of mesic and wet forest with 4 diamond-shaped ½ inch long leaves at each node and typical peperomia flower spikes. The leaves are usually succulent.

have a reddish tinge and in some the underside of the leaves may be red or purple. I will describe 2 of the most widespread species in the hope that once you are familiar with these, you will have no difficulty in recognizing the rest as close relatives.

P. tetraphylla is a common indigenous plant of mesic to wet forest and even alpine deserts. It is a very hardy plant that has a remarkable ability to grow on what appears to be an almost bare rock, and will spread its succulent low mats across several square feet in favorable habitats. Four leaves, about ½ inch long usually sprout from a common node on the stem. These leaves are roughly diamond-shaped and succulent. The plant bears typical elongated cylindrical flowering spikes. This plant is found throughout the tropics, and on all the major Islands in Hawai'i.

P. remyi is an endemic native with dark reddish-purple branching stems and short-stemmed pointed oval leaves about 1 or 2 inches long. The plant may be erect or sprawling. The leaves are a

Look for: A succulent plant of fairly moist areas with opposite oval pointed leaves, reddish stems, and typical peperomia fruiting spikes.

dark yellowish-green on top but the lower surface is paler and often has reddish veins or a reddish tinge. Several long slender flower spikes, up to 6 inches long, bear the inconspicuous flowers and fruit. This plant is found in mesic valleys and wet forest on all the larger Islands.

Pre-contact Hawaiians incorporated parts of unspecified 'ala'ala wai nui species into medicinal mixtures intended to treat conditions as varied as irregular menstrual periods, chest pain, thrush, vaginal discharge, asthma, uterine abnormalities, and tuberculosis. A gray-green dye was also obtained from these plants.

'AWAPUHI, SHAMPOO GINGER
Zingiber zerumbet
Zingiberaceae (ornamental gingers, cooking ginger)

This plant has a long history of cultivation in southeast Asia and was spread throughout the islands of the Pacific by early voyagers. It was introduced into Hawai'i in prehistoric times by the Polynesian settlers and is now common in shady valleys in disturbed, mesic sites. As in the other gingers we discuss, the unbranched stalk rises directly from underground stems or rhizomes. The plant is commonly only about 2 feet high with alternating lance-shaped leaves about 8 to 18 inches long and 1 to 3 inches wide. A separate, leafless stalk bears the flowering structure, which is an egg-shaped body with overlapping scales that looks vaguely like a pine cone. This is up to 4 inches long and 2 or 3 wide and usually turns red when mature. White or yellowish flowers emerge from between the scales. When squeezed, this

Look for: A low growing plant of shady valleys with a single stem bearing large, blade-like leaves mixed with stalks carrying scaly oval red bodies that yield a slimy liquid when squeezed.

flowering head yields copious quantities of a slimy clear fragrant liquid that the Hawaiians used as shampoo, giving the plant its English common name. The plant is seasonal and dies down during the cooler months.

Hawaiians had a number of uses for this plant. Ashes of the leaves were a component in a poultice used to treat cuts and bruises. Sliced, dried, and powdered rhizomes (which resemble the familiar cooking ginger) were part of mixtures used to relieve headaches, toothache, ringworm and other skin diseases, aching joints, indigestion, sprains, and stuffy noses and sinus congestion, as well as growths in the nose. Powdered rhizome was also used to scent kapa cloth and to flavor meat.

BAMBOO ORCHID
Arundina graminifolia
Orchidaceae (orchids)

This orchid is native to south and south east Asia and was introduced to Hawai'i, probably as a garden ornamental, sometime before 1945 when it was first collected in the wild. Unlike many orchids which do not set seeds here due to the absence of suitable pollinators, this plant produces large numbers of tiny, wind-borne seeds which spread readily. It does not occur in such dense stands that it crowds out other plants, and seems unlikely to be a problem in our ecosystem, however. The orchid grows on long slender stalks, often 5 feet in height, which bear one or two flowers at a time. The alternate leaves are well spaced along the stalk and are long and slender and come to a sharp point. The flower has the typical orchid form, with 3 prominent, and 2 less obvious, white or pinkish petals forming a flat backdrop for an attractive reddish-purple funnel with a yellow throat. Flowers are about 2 inches across. A cylindrical ridged seed capsule about 2 inches long often follows the flowers.

KEN SUZUKI

There are only 3 species of orchids that are endemic to Hawai'i and these are so rare and so inconspicuous that you are unlikely to see one unless you hike with a knowledge-

Look for: A slender, erect, reed-like plant with typical orchid flowers colored a pale pink with a reddish-purple funnel and yellow throat.

able botanist who can show you where they may be found. While many orchids are epiphytes, meaning that they grow on another plant, most of those you will see here are ground dwellers. This includes the native species, which have suffered severely from the trampling and rooting of pigs. Orchids have

73

very tiny seeds which are readily spread by the wind. Since many species of these plants are found in regions with a climate similar to that of Hawai'i, it might seem strange that so few were able to become established here. The tiny seeds often need the help of a compatible fungus to develop into a plant however, and the absence of the right fungus, combined with the lack of suitable pollinators may have limited the success of these plants.

BASKET GRASS
Oplismenus hirtellus
Poaceae (bamboo, grasses)

This grass is a common ground cover in shady mesic forests. It is a delicate, much branched creeping perennial grass with lance-shaped alternate leaves widely spaced along a thin stem. The leaves are 2 to 4 inches long and ½ inch wide and often somewhat hairy. New leaves emerge from the tip of the stalk as long, slender, conical spikes. In plants growing in

Look for: A dry, thin stalked sprawling grass with well spaced lance-shaped alternate often rippled leaves, covering the ground in shady valleys and slopes in mesic forests.

moist conditions, the leaves are not perfectly flat, but have small, regular, ripples on the surface. In form, it resembles honohono grass, but does not look nearly as lush and succulent, being dry and more slender in the stalk. Honohono grass may also have a ripple or two on the leaf surface, but these are not as frequent or regular as in basket grass. The plant has been called "honohono kukui" or "honohono maoli" among other things because of their similarities, however. Basket grass was introduced to the Islands very soon after Western contact, being first reported here in 1819. It is a native of tropical America.

One form of basket grass has pink and white striped leaves. This has been used as a garden ornamental and is sometimes planted in hanging baskets, which may explain the common name.

CHINESE GROUND ORCHID, NUN'S ORCHID, NUN'S HOOD ORCHID
Phaius tankarvilleae
Orchidaceae (orchids)

This orchid was introduced to the Islands as an ornamental and proved capable of spreading into the wild. It was first reported on our trails about 1931. The plant is native to southern China and the Pacific islands north of Australia. It grows in shady areas in disturbed mesic and wet forest in Hawai'i. The foliage of this orchid rises from the base in clusters of pleated green leaves, about 2 feet long, narrow at the base and tapering to a point. Flowers are borne on a leafless stalk up to 3 feet tall. They are large, about 3 inches in diameter, with a typical orchid shape, white outside, and a dull brownish-purple inside.

KEN SUZUKI

Look for: A conspicuous cluster of orchid flowers on a 2 to 3 foot stalk, each flower 3 inches in diameter with a white and brownish-purple color scheme.

CHINESE VIOLET
Asystasia gangetica
Acanthaceae (white shrimp plant, thunbergia)

This plant is a sprawling, viny herb that extends 1 to 6 feet in length. The 1 to 2 inch long leaves are heart-shaped or pointed ovals in form, and are paired, opposite, and well-spaced along the slender stem. The flower buds tend to lie in a row along one side of a spike and open to reveal purple, or sometimes white, 5-petalled blossoms about 1 inch in diameter. The flower often appears to be slightly rectangular, being a little wider in

Look for: A sprawling, viny herb with purple (or white) flowers about 1 inch in diameter in lowland mesic environments.

the horizontal dimension than in the vertical. The plant is native to Africa, India, and the Malay Peninsula and is widely grown in the tropics as a filler and ground cover. It was first noticed in the wild on O'ahu about 1925, and is found naturalized near urban areas in lowland disturbed mesic habitats.

ELEPHANTOPUS, ELEPHANT'S FOOT
Elephantopus mollis
Asteraceae (aster, ragweed, nehe, lettuce)

This is a coarse, weedy herb with soft, usually somewhat furry pointed-oval leaves about 3 to 8 inches long, with prominent veins and small teeth along the edges. Young plants are low-growing rosette-like herbs but send up slender, branching stalks with sparse alternate leaves when they are ready to flower. The stalks may be 2 or 3 feet tall, and each branch-tip bears a tight cluster of spiky looking fruiting bodies on top of a flat platter made of 2 or 3 small triangular leaf-like structures. This introduced weed is probably native to tropical America, but is now wide spread throughout the tropics. It was first reported in Hawai'i in 1926, and is found in dry pastures and along roadsides and trails in mesic and wet forest on all the larger Islands.

Look for: A low growing, coarse, rosette-like weed that sends up a slender, branching stalk bearing tight clusters of spiky fruiting bodies set on collars of triangular leaf-like structures.

A second species, *E. spicatus*, also common in Hawai'i. It is an erect plant with smaller leaves and with spindle-shaped, rather than triangular, leaf-like appendages below the flower heads. It favors somewhat drier areas. This plant is native to Latin America and was first collected on O'ahu in 1935.

FALSE MEADOW BEAUTY
Pterolepis glomerata
Melastomataceae (clidemia, tibouchina, miconia)

This herb is native to tropical eastern South America. It was first collected on O'ahu in 1949 where it is found in upper mesic to wet disturbed areas, especially along the edges of trails. It is a low growing plant,

Look for: A low growing trail-side herb with pointed oval leaves that have 3 veins running from base to tip and pale lavender 4 petalled flowers.

JOHN HOOVER

rarely more than a foot or so high. The leaves are pointed ovals about 1 inch long, with 3 prominent veins that run from the base to the tip in typical melastome fashion. The flowers have 4 pale pink or lavender petals and are found at or near the tip of the stem.

FIREWEED
Erechtites valerianifolia
Asteraceae (goldenrod, sagebrush, verbesina)

This plant is an erect, usually un-branched herb up to 6 feet tall with alternate, deeply lobed and highly serrated leaves. Numerous small flower heads on slender branching stems are borne at the top of the plant. These are followed by small, wind-borne seeds. This weed is na-tive to Latin America and was first collected in Hawai'i in 1916. It is found in relatively wet forest. A rela-tive, *E. hieracifolia*, is common in drier areas of the Islands, up to about 1000 feet of elevation.

Look for: A tall, straight weedy plant with alternate deeply lobed and serrated leaves in mesic to wet forest.

None of my references states that the leaves of this plant are edible as a salad green, although one does say that the young inflorescences of the lowland species are a popular food in Java. When I was first introduced to the plant I was told that it could be eaten as a kind of "lettuce" and have found it quite palatable as a raw green. Always remember, of course, that you should never eat any strange plant unless you are quite certain that it is safe, and then it is probably best to try a small sample and wait to see if it agrees with you before indulging heavily.

GUINEA GRASS
Panicum maximum
Poaceae (grasses, bamboo, sugar cane, and all the major food grains)

You can not overlook this ob-noxious grass when it chokes our trails. It is a robust plant that grows to be 9 feet tall and is often covered with mod-erately irritating hairs. This grass was introduced from Africa in about 1871 as cattle forage, but has recently begun to spread along our trails in the upper dry to wet forest. It

Look for: A large grass growing higher than your head that forms dense stands difficult to penetrate.

has become a serious pest only in the last 5 or 10 years. It forms dense stands over wide areas and can completely obscure a trail. Even when the route is obvious, pushing through a healthy thicket of this grass can be difficult.

Both guinea grass and palm grass produce an abundance of tiny seeds that will cling to your clothes, boots, and gear. Please be careful to clean these off before you enter a different trail to avoid spreading these weeds into any area that they may not already infest.

HONOHONO
Commelina spp.
Commelinaceae (wandering Jew, spiderwort)

Two species of Commelina have been in-troduced into Hawai'i. Both come from the Old World tropics and have been here for many years. *C. diffusa* was first found here about 1837, and *C. benghalensis* in 1909. These sprawling, grass-like herbs often form the ground cover in shady moist areas in mesic valleys and disturbed areas in wet forests. They have lance-like leaves about 2 to 3 inches long that emerge in alternation from nodes along the stem. The plants readily root at the nodes. You will sometimes see small blue flowers on the herb. There are 3 petals, 1 of them being much smaller than the other two, which are round, and give the blossom the appearance of a mickey mouse face, with greatly exaggerated ears. These are tender-looking plants and are relished by cattle. Reportedly, *C. diffusa* can also be eaten, raw or cooked, by people, although I have never heard of anyone doing this. The plant resembles basket grass, which is drier with a more slender stalk and never has blue flowers.

Look for: A sprawling, tender ground cover in shady valleys in intermediate forest with well separated lance-shaped alternate leaves and sometimes bright blue flowers with 3 petals, one smaller than the others.

HULUMOA
Korthalsella complanata
Viscaceae (mistletoe)

There are 4 endemic and 2 indigenous species in this genus in Hawai'i. All are virtually leafless parasitic herbs that grow on the branches of woody plants and obtain part of their nourishment from them. The indigenous species above is the most common and the one you are most likely to see along our trails. It has jointed flattened green branching stems that are about 4 to 10 inches long. Each segmented stem somewhat resembles a length of tapeworm. The plant is generally erect. Tiny, inconspicuous leaves, flowers, and fruit may be attached to the stems at the joints. This native mistletoe is often seen growing on ʻōhiʻa ha, koa, or ʻōhiʻa trees, but may occur on a wide variety of other native trees and shrubs as well. It is fairly common in mesic and wet forest on all of the major Islands. This species occurs on Henderson Island also. The other species are similar, although most of the endemic plants have cylindrical rather than flattened stems and some of them droop. "Hulumoa" means "chicken feathers".

Look for: Jointed branching green flattened or cylindrical stems growing on the woody branches of native trees or shrubs in mesic or wet forest.

MINTS
Phyllostegia spp., Stenogyne spp.
Lamiaceae (thyme, catnip, pennyroyal, rosemary)

Many endemic species of native mints are found in these 2 genera which are very similar in overall appearance, being generally rather sprawling, vine-like herbs or sub-shrubs. Unfortunately, many of the species are thought to be extinct and most are rare and confined to very localized habitats. Most, but not all, of the species have the quadrangular stems that are common in this family. None of the natives have the highly aromatic oils that are so characteristic of many of their relatives, and which probably repel both mammalian and insect herbivores by virtue of their bitter taste. Either the plants lost their repellant oils after arriving in Hawaiʻi, since they were no longer needed in the absence these plant-eaters, or the original colonizing ancestor did not have them in the first place. I will describe one species from each genus, in the belief that once you can recognize these, you will also be able to identify its relatives as being mints, at least.

Phyllostegia glabra is an erect to sprawling herb with square stems and opposite oval leaves that taper to a sharp point and are 3 to 9 inches long. The leaves often have a reddish tinge or red veins, and have toothed edges. The white flowers are about an inch long, sometimes have a touch of purple in them, and occur in open spikes. As in most members of this genus, the lower lip of the flower is larger than the upper, which distinguishes these

79

plants from those in *Stenogyne*. The fruit is a small dark green to black berry. The plants can be found in mesic to wet forest on Oʻahu, Molokaʻi, Lānaʻi, and East Maui. Almost the entire genus is endemic to Hawaiʻi, although 1 species is found in Tahiti. Twenty seven species were described from the Islands, but 7 of these seem to be extinct and 5 are rare. None are really common.

Look for: A sprawling herb with square stems, opposite spearhead shaped leaves with teeth, and whitish flowers with the lower lip larger than the upper.

KEN SUZUKI

Stenogyne kaalae is a viny herb with square stems up to 6 feet long. The opposite leaves are pointed ellipses with serrated edges and about 1 ½ to 3 inches long. The small flowers are a dark maroon in color, and as is usual for species in this genus, the upper lip is larger than the lower, a characteristic that separates these plants from those in the genus above. The flowers are ½ inch or less in length. The small berries are bluish-black when ripe. This species

Look for: A viny herb with square stems, opposite pointed leaves with serrated edges and dark maroon flowers with a protruding upper lip.

can be seen in mesic forest in the Waiʻanae Mountains on Oʻahu. The genus is endemic to Hawaiʻi where 20 species were originally described. Four of these are now extinct and 5 more endangered. None are very common.

MONTBRETIA
Crocosmia x crocosmiiflora
Iridaceae (iris)

This plant, whose ancestors come from eastern South Africa, has been in Hawaiʻi for about 100 years, having been found in the wild on Oʻahu in 1909. It was almost certainly introduced as a garden plant and escaped from cultivation or was planted in the mountains by someone who admired its blossoms. Montbretia, or day lily as it is sometimes called, is a hybrid of two different species, *Crocosmia pottsii* and *C. aurea*, and like many hybrids, it does not produce fertile seeds but spreads aggressively by means of underground stolons and corms. It has done this successfully enough that it is not uncommonly

found growing wild along roadsides and trails in mesic to wet forest. The plant has typical lily-like long, narrow leaves, 12 to 18 inches in length and 1 inch or less wide with a central mid-rib. The most striking feature of the plant is the slender branching zig-zag stem that bears colorful orange flowers about 1 ½ inches in diameter. These have 6 identical lily-like petals and begin to open from the bottom of the cluster, one or a few at a time. The younger buds alternate on opposite sides of the stalk above the blossoms.

Look for: A lily-like plant with bright orange 6-petalled flowers about 1 ½ inches across.

PALM GRASS
Setaria palmifolia
Poaceae (grasses, bamboo, sugar cane, and all the major food grains)

This aggressive grass grows to be 1 to 3 feet tall with broad, finely pleated, dark green leaves. The leaves are about 1 to 2 feet long and 2 or 3 inches wide. They resemble those of the Philippine ground orchid, but are less rigid, and with finer, more regular pleats. Palm grass leaves are extremely abrasive when stroked from tip to base. Slender seed stalks with open, scattered branches rise above the leaves. This iniquitous plant is native to tropical Asia and was first reported from Hawai'i about 1903. It is now widely naturalized in mesic and wet forest, where it often

Look for: A 1 to 3 foot tall grass with broad, finely pleated leaves and tall, open, flowering stalks.

obscures footing on our trails and requires increased effort to walk through the dense foliage. Like guinea grass, this plant has only become widespread within the past few years.

Both guinea grass and palm grass produce an abundance of tiny seeds that will cling to your clothes and boots. Please be careful to clean these off before you walk on a different trail to avoid spreading the plants into any area that they may not already infest. Unfortunately, it is not practical to clean boots and clothes thoroughly after passing through every patch of these grasses, so once they become established near a trail head, it seems inevitable that they will continue to migrate up that trail until reaching their habitat limit. They form dense stands and will replace more diverse and interesting native species where they become established.

PHILIPPINE GROUND ORCHID
Spathoglottis plicata
Orchidaceae (orchids)

KEN SUZUKI

This orchid is native to southeast Asia and adjacent islands. It was introduced into Hawai'i by Harold Lyon, founder of the Lyon Arboretum at the head of Manoa valley, and soon became naturalized in the wild, where it was reported by 1929. The orchid is found scattered throughout our mountains in disturbed mesic and wet forest areas. The 2 foot-long leaves arise from the base of the plant and are broad, stiff, conspicuously ridged and spindle-shaped. There is no obvious midrib. They resemble the leaves of palm grass, but are not hairy or abrasive. Pink to maroon flowers about 1 inch in diameter are borne on sturdy

Look for: A cluster of markedly pleated, spindle-shaped leaves rising from the ground with one or more pink to maroon, orchid-shaped flowers about 1 inch in diameter near the end of a stalk.

leafless stalks about 2 feet long. They have a typical orchid form but are uniform in color unlike the other orchids mentioned here. Cylindrical, ribbed seed pods about 2 inches long appear below the flowers.

RATTLEPOD
Crotalaria spp.
Fabaceae (acacia, garbanzo bean, clover)

Eleven species in this genus are naturalized as weeds in Hawai'i. These are herbs or shrubs with alternate leaves that are simple or compound with 3 leaflets. The pea-like flowers are yellow, sometimes with streaks of other colors, and are borne in narrow clusters at the stem tips or in leaf axils (where the leaf joins the stem). The rattlepod fruit is an inflated bean pod in which the ripe seeds will rattle when shaken. Different species are native to tropical areas from around the world—Africa, India, Asia, and the America's. They have arrived in Hawai'i at various times, beginning in about 1864 and extending through the 1980s, and are now naturalized along trails, roadsides, and pastures in mesic areas. Some may have been introduced for ground cover, green manure, or as ornamentals.

Look for: Weedy herbs or shrubs with yellow, pea-like flowers, and inflated pods that rattle when shaken.

RUMEX, HUʻAHUʻAKŌ
Rumex albescens
Polygonaceae (buckwheat, sea grape, Mexican creeper, rhubarb)

There are 7 species in this genus in Hawaiʻi, 3 endemic and the rest introduced aliens. I will describe the one above, with which I am familiar, and trust that you will be able to recognize its native relatives at least, if you encounter them. Huʻahuʻakō or rumex, is a large sprawling thick-stemmed ungainly viny herb with large alternate lance-shaped leaves about 6 to 12 inches long. The plant is light green in color. In season, the female plant bears open clusters of small, 3-flanged, pointed yellow-green flowers with no obvious throats, which are followed by small brown nut-like fruits. This rumex is found scattered in mesic forest on Kauaʻi and in the Waiʻanae Mountains on Oʻahu.

KEN SUZUKI

R. giganteus is a larger native species found on Maui, Molokaʻi, and Hawaiʻi, while pāwale or *R. skottsbergii*, is a native shrub of dry areas on Hawaiʻi. I have found no mention of any use made of these plants by the early Hawaiians.

Look for: A large, sprawling herb with alternate large, lance-shaped leaves in mesic forest on Kauaʻi or the Waiʻanae Mountains on Oʻahu.

SEDGES
Carex spp., Gahnia spp.
Cyperaceae (ʻuki, papyrus, umbrella plant)

Besides the two species of ʻuki (Machaerina) described in this book, there are many other sedges along our trails in Hawaiʻi. Some of these are endemic while others are indigenous or introduced species. Sedges are plants that usually have long, slender strap-like stems that taper to slender points much like the leaves of the closely related grasses. Sedges differ from grasses in that the stem is usually triangular in cross section and solid, while grasses have round stems with many nodes, and often the stem between two nodes is hollow. It may be necessary to feel around the base of the plant to detect the triangular stem of a sedge, however. The 2 sedges that I will discuss form dense clumps of grass-like blades that arise from the base of the plant. The 2 species are readily separated when in fruit, but can be difficult to tell apart when no reproductive structures are present. The Hawaiians called sedges mauʻu or kāluhāluhā, but I will use the English name.

Carex wahuensis is a common endemic sedge found in dry to mesic forest and in subalpine scrublands on all the larger Islands. The leaf blades are up to 3 feet long. When in fruit, it produces a stem bearing a few long slender cylindrical spikes that are broad at the base, where the bristly pointed oval seeds develop, and then taper to a conical point.

JOHN HOOVER

Look for: Grass-like clumps of slender strap-like blades that taper to sharp points, are about 1 to 4 feet tall and have triangular stems at the base.

Gahnia beecheyi is another common endemic that occurs in mesic and sometimes wet forest on most of the larger Islands. The leaf blades are up to 4 feet long. Its fruiting body is more open and shaggy with similar bristly pointed oval seeds arranged in large rather shapeless clusters on top of a stem.

The Hawaiians used 2 sedges for making mats. The common bulrush or 'aka'akai (*Schoenopectus lacustris*) could be used for easily made but not very durable mats that were often used as pads on the pebble floor to protect the more valuable lau hala mats from abrasion. A finer sedge, makaloa (*Cyperus laevigatus*) was woven into very fine, beautifully decorated mats, perhaps the finest in Polynesia, that were used for the sleeping mats of the ali'i. These were most often produced on Ni'ihau, which was famous for its fine makaloa mats.

'UKI'UKI
Dianella sandwicensis
Liliaceae (now assigned to Phormiaceae?) (lilies, amaryllis)

This indigenous lily, which is also found in the Marquesas, is most common in the mesic forest, but may be found in dry shrub-lands or wet forests also. The typical smooth, glossy strap leaves rise in alternation from the base and can be distinguished from a common

Thomas H. Rau

sedge (also called ʻuki) by the presence of a prominent central vein running the length of the leaf, which the sedge does not have. When flowering, the plant produces a stalk bearing many small, 6 petalled, lily-like flowers about ½ inch in diameter. These are pale blue to white in color. The flowers are followed by strikingly beautiful blue to purplish ½ inch-long berries. A joy to behold!

The Hawaiians used the fruits to prepare a blue or blue-purple dye, or strung them with added greenery to make an attractive lei. The leaves or fiber obtained from them could be braided to make cords which were sometimes used to secure thatch to the houses or for other purposes. Where abundant, the plant may have been used as thatch.

Look for: Typical lily-like strap leaves with a midrib, small bluish or white 6 petalled flowers, and lovely blue berries.

WEDELIA
Wedelia trilobata
Asteraceae (sunflower, daisy, thistle)

This is a creeping, mat-forming herb with opposite coarse toothed leaves about 1 ½ to 4 inches long. The leaves have little or no stems and often appear to clasp the stalk of the plant. Wedelia tends to spread by rooting at the nodes. The plant bears numerous small yellow sunflower-like blossoms individually at the tip of each stem. These have about 10 bright yellow petals with small teeth at the tips, and yellow centers, and

Look for: A mat-forming herb with opposite, toothed leaves and many individual yellow sunflower-like blossoms, about 1 inch across.

are an inch or a bit more across. The plant is native to the American tropics and is extensively cultivated in Hawaiʻi as a ground cover. We see it growing wild occasionally in mesic to wet forest and it has the potential to become a serious pest if it is able to spread widely since it can grow over and smother other low growing plants.

WHITE SHRIMP PLANT
Justicia betonica
Acanthaceae (Chinese violet, thunbergia)

This plant is a coarse herb or weak-stemmed shrub, about 1 ½ to 8 feet tall with purple-tinged stems. The leaves are pointed ovals, 3 to 10 inches long by 1 ½ to 6 inches wide and have scanty hairs along the edges and the veins. They are opposite and well-spaced along the stem. The flowers are tightly clustered in 4-inch long erect spikes at the ends of the branches. The most conspicuous feature of the flower cluster is the set of 3 prominent white leaflets with marked green veins that surround each flower and overlap along the spike. With a certain effort of the imagination, this spike can be seen as the tail of a cooked, husked, green-veined shrimp. The plant is native to tropical Asia and was introduced to Hawai'i as an ornamental, being first collected in the wild on the Big Island in 1943. It often escapes from gardens to form extensive stands in disturbed, mesic environments near towns, and appears to have the potential of becoming a serious weed pest.

Look for: A rather weedy, erect herb bearing 4-inch long spikes composed of overlapping white, green-veined leaflets that vaguely resemble a shrimp's tail.

SHRUBS

CLIDEMIA, KOSTER'S CURSE
Clidemia hirta
Melastomataceae (Miconia, Tibouchina)

When I first began hiking in Hawai'i in 1962, we would occasionally encounter clidemia, but it was not yet widespread. Now it is everywhere, and is undoubtedly the most widespread, obnoxious, and invasive of the shrubby alien weeds. Clidemia is native to the American tropics and was first reported wild in Hawai'i about 1941. We are not sure how or why it was brought in. The shrub can be found from dry forest to the wet forest. It is sometimes called Koster's curse, perhaps in "honor" of the person who introduced it to the Islands? But most people call it clidemia. You will commonly encounter it as a low, 2 or 3 foot high shrub, although in favorable locations it will get to be much larger, nearly to the size of a small tree, and form dense thickets that crowd out all other vegetation and make trails nearly impassible. Native mosses, ferns, and liverworts that help form a soil-binding mat die off under the thicket of clidemia, reducing the water-holding capacity of the soil and increasing the likelihood of erosion.

The melastomes are easily recognized since their leaves are rather different from those of most other plants. In addition to the midrib found in most leaves, melastomes have other prominent longitudinal ribs that run from the base to the tip of the leaf. This can easily be seen in clidemia, in which the hairy leaf has a tear-drop shape, with the stem at the round end. In addition to the midrib, a pair of obvious ribs run parallel to the edge of the leaf, and near it, and a second pair runs between these and the midrib, so that there are 5 prominent veins extending from the stem to the tip of the leaf. Transverse veins join these main ribs, dividing the leaf into a serious of rectangles. The veins are recessed on the top of the leaf, and protruding from the bottom, so that from the top, these

JOHN HOOVER

Look for: A shrub with hairy, tear-drop shaped leaves with 5 prominent longitudinal veins, and slightly elevated rectangles of leaf surface between them.

87

rectangles rise above them and give the leaf a somewhat crinkled appearance. The leaves are arranged in opposite pairs along the stem. The plant produces white 5-petaled flowers about ½ inch across, followed by small, clustered, purple fruits that can be eaten, although I do not find them very tasty. The birds seem to like them though, and this has greatly accelerated the spread of this noxious weed.

DOG TAIL
Buddleia asiatica
Buddleiaceae (butterfly bush, smoke bush)

This weedy shrub is an ugly, scraggly bush with opposite lance-shaped leaves about 2 to 6 inches long. It bears tiny white fragrant flowers on long slender tail-like spikes, 3 to 9 inches long, which usually bear small brown seed capsules in addition to the flowers. The plant is native to south Asia and nearby islands and is now naturalized in Hawai'i in mesic and wet forest. It was first collected here in 1908 and has become a common weed.

Look for: A scraggly, open shrub with slender, lance-shaped leaves and long tapering spikes of tiny white fragrant flowers.

HĀHĀ, CYANEA
Cyanea angustifolia
Campanulaceae (lobelias, bellflowers)

On September 10, 1911, the Sunday Advertiser had an article on a new trail that had just been constructed starting in Nu'uanu Valley on O'ahu. In mentioning sights worth the notice of photographers, it stated, "The lobelia, of which there are a hundred varieties in the island, stands out conspicuously." Alas, thanks to the depredations of rats, pigs, and other feral animals, the attacks on the plants by slugs and insects, and the loss of the lovely native honey creeper and honey eater birds that once pollinated them, the diverse members of the lobelia family that once graced our mountains are no longer conspicuous. Of the original 52 species of *Cyanea* that have been described, 20 are probably extinct and many others are very uncommon. Because they are now so rare, I will discuss only the two groups that you are most likely to see, with the plant above representing one once large and highly varied clan. See 'Ohā wai, *Clermontia*, for a discussion of the other.

Most of the *Cyanea,* or hāhā, like their relatives in the genera *Brighamia, Delissea, Lobelia, Rollandia,* and *Trematolobelia,* resemble small palm trees. They very greatly in size, from just a foot or so in height to giants towering nearly 40 feet into the air. Their leaves are clustered at the top of the stalk, and generally are fairly large pointed oblongs, although some will be narrow and strap-like while in other species the leaves may be fern-like with intricately scalloped lobes and even prickles. Most have a single, somewhat fleshy stem, or branch very sparingly, but a few, including the one above, may branch and become a small tree very like the ʻohā wai. As in ʻohā wai, a regular pattern of prominent round leaf scars can be seen on the stem, especially just below the

tuft of leaves. I will describe the species above, as being the one you are most likely to see, as it is fairly common on Oʻahu, and also occurs in mesic and wet forests on Molokaʻi, Lānaʻi, and West Maui, whereas most species are confined to a single Island. Once you can recognize this one, you will probably be able to identify the others as its relatives.

KEN SUZUKI

Cyanea angustifolia is a shrub or small tree 4 to 16 feet tall with lance-shaped leaves 4 to 12 inches long clustered at the tips of the branches. You are most likely to see unbranched or sparsely branched young plants that look like miniature palms, although on rare occasions I have run across 12-foot tall branching trees of this species. Many small flowers cluster on drooping slender stalks that hang down below the leaves. In this species these will be white to greenish-white, possibly with a tinge of pale purple. The flowers are followed by small purple berries.

Look for: A small palm-like plant with a tuft of pointed oval leaves at the top of a stalk in mesic to wet forests on Oʻahu and the 3 younger Islands of Maui Nui to the east.

Other species in this group have a variety of flower colors and flowering stalks. Flowers may be blue, pink, red, white, green, or yellowish with some bright and gaudy and others subdued and delicately tinged or striped. *Lobelia* bears its flowers in a single vertical spike, while *Trematolobelia,* which also projects its flowers above the tuft of leaves, sends out 3 or 4 horizontal arms, each 1 or 2 feet long, like the spokes of a wheel, on which the white or pink blossoms appear. This plant produces dry little pepper-shaker seed capsules instead of the orange or purple juicy fruits found in most other genera. Plants in both of these 2 genera die after blooming, unlike hāhā and members of the remaining genera which can bloom repeatedly. The Hawaiians on the Big Island are reported to have cooked and eaten the leaves of one species of hāhā, *Cyanea platyphylla,* which they called ʻakūʻakū.

HŌʻAWA

Pittosporum spp.

Pittosporaceae (I know of no familiar relatives)

There are 10 endemic species of
hōʻawa in Hawaiʻi and 2 introduced
ones. They occur in mesic and wet forest, and rarely in very dry areas. The
plants are shrubs or small trees, usually rather gangly in form, with 4 to
10 inch long leaves clustered near the
tips of the branches. The edges of the
leaves tend to curl under a bit, and in
many species the underside is covered
with light brown hairs, especially in
new leaves. Small clusters of white or
cream-colored flowers are sometimes
seen, usually on the stem below the
leaves. The flowers are in the form of

JOHN HOOVER

Look for: A gangly shrub or small
tree with simple, alternate leaves
clustering near the branch tips. The
leaves are 4 to 10 inches long, with
edges that curl under slightly, and
often hairy. The fruit resembles a
walnut.

short tubes with about 5 petals and are ¾ inch long. These are followed by
a large fruit, about the size and shape of a walnut, with a wrinkled surface
that is divided into 2 or 4 equal segments by narrow grooves. These often
occur in pairs, but may be single or in clusters. When the fruit splits open, it
exposes the gummy orange pulp and shiny black seeds. These were one of the
favored foods of the native Hawaiian crow, or ʻalalā, which is now extinct in
the wild. The light-colored wood was sometimes used by early Hawaiians for
the gunwales of canoes.

ILIAU

Wilkesia gymnoxiphium

Asteraceae (aster, cosmos, dahlia)

This curious endemic is an erect,
rarely branching plant that grows to
be 3 to 15 feet in height and bears
a tuft of narrow, sword-like leaves 6
to 20 inches long and ½ inch wide
on the top of a long slender stalk.
A number of parallel veins run the
length of each leaf. Large spikes
of cream-colored flowers about an
inch in diameter with no petals may
emerge at the end of the stalk during

VINCENT T. SOEDA

Look for: A small Dr. Seuss kind of tree
with a tuft of thin, strap-like leaves at
the end of a long stalk in the Waimea
Canyon - Kōkeʻe area on Kauaʻi

the flowering season from May to July. As many as 100 blossoms may be produced by a healthy specimen, but the plant only blooms once and then dies. This plant is found only on dry ridges and open mesic forest in the vicinity of Waimea Canyon and ridges to the west of it on Kaua'i, but is readily seen on the Iliau Loop Trail on the Canyon rim, or near the end of the 'Awa'awapuhi Trail starting further up the road in Kōke'e. A much smaller, rarer, and less accessible relative, *W. hobdyi*, also occurs in this area.

KOKI'O KE'OKE'O, HIBISCUS
Hibiscus arnottianus
Malvaceae ('ilima, ma'o, okra, hau)

The *Manual* lists 10 species in this genus in Hawai'i, 5 of them endemic, 3 naturalized aliens, 1 indigenous, and 1 that may be indigenous or a Polynesian introduction. Several of the endemic species have red or yellow flowers, including our yellow-flowered *H. brackenridgei* which is the State Flower of Hawai'i. These red and yellow flowered species, and one white flowered one, are rare and you are unlikely to encounter them in the wild, so I will describe the one above, which is relatively easily seen. Koki'o ke'oke'o is a shrub or small tree that grows up to 30 feet tall. Some on Mt. Tantalus on O'ahu are quite large, sending up numerous sturdy trunks from a base 2 feet or more in diameter. The leaves are broad ovals tapering to blunt points at each end and are 2 to 6 inches long by 1 to 4 inches wide. The large white blossoms, 4 or 5 inches across, have 5 petals and are somewhat fragrant. A bright pink column bearing the stamens adds a colorful note to the center of the flower. These plants occur in mesic and wet forest on O'ahu and in Wailau and Pelekunu Valleys on Moloka'i.

KEN SUZUKI

Look for: A sizable shrub of mesic or wet forest on O'ahu or Moloka'i with large, 5-petaled, white flowers with long, straight pink columns emerging from the centers.

Hawaiian hibiscus have been used extensively in hybridization experiments to create a wide variety of attractive horticultural varieties. Since certain native species, especially *H. waimeae* from Kaua'i, have a pleasant fragrance, a trait which is rare in this family, these plants have been used to introduce an attractive odor into garden cultivars. In old Hawai'i, the buds and leaves of hibiscus were chewed to relieve constipation, and the roots were included in a concoction designed to "purify the blood". Pink and lavender dyes could be obtained from petals of the flowers, and the flowers themselves were used for adornment.

LABORDIA, KĀMAKAHALA
Labordia spp.
Loganiaceae (plants yielding strychnine, curare; pua kenikeni)

JOHN HOOVER

Look for: Shrubs or small trees of mesic or wet forest with opposite leaves that clasp the twig and which have veins that appear translucent when viewed against the sky.

Labordia or kāmakahala? This is a gray area in that we have a group of plants that were named by the Hawaiians but apparently played no significant role in their culture and so are now known mostly to botanists who are more likely to use the scientific name. Since in this case the scientific name is also shorter and simpler than the Hawaiian, you might prefer to learn it. There are 15 endemic species in this genus in Hawai'i, many of them rare, and most found in very limited areas on a single Island. The plants are shrubs or small trees with opposite leaves that are pale green beneath and usually have a small leaf-like structure where the leaf stem meets the twig, that seems to clasp the twig. The flowers are usually tubular, opening into 5 or 6 lobes, and are greenish-yellow, yellow, orangish-yellow or rarely white in color. Labordia are found in mesic to wet forests. I will describe the species that you are most likely to encounter, *L. tinifolia*.

L. tinifolia is a small tree with light green, pointed elliptical leaves that are about 2 to 8 inches long and 1 to 2 inches wide. The base of the paired leaf stems forms a sheath around the twig. As in other labordia, if you look through a leaf at the sky, you can see that the veins are translucent, and appear lighter than the rest of the leaf. This is a characteristic that I find helpful in recognizing these plants. This small tree is found in mesic to wet forest on all the major Islands.

Several low-growing species of labordia occur on the wet, windswept summits of mountains or ridges and have small but very lovely tubular yellow flowers, well worth noticing when you are exploring these areas. Seed capsules of some labordia were used in lei by the native Hawaiian peoples, but I know of no other uses that they had for these plants.

MĀMAKI
Pipturus albidus
Urticaceae (olonā, artillery plant, nettle)

Māmaki is an endemic shrub or small tree, up to 20 feet tall, with coarse alternate broadly oval leaves that are often as wide as they are long, come to a sharp point, and have small teeth along each edge. They are 4 to 10 inches long with 3 large veins radiating from the base. The leaves are green above and often pale, almost white, with fine hairs, beneath. The veins often have a reddish tinge. Young branches are covered with a mat of gray, wooly hairs.

Inconspicuous flowers hug the stem and are followed by whitish dry insipid, somewhat fleshy, mulberry-like fruits. The plant is scattered in mesic and wet forest, especially in gulches, on all the main Islands.

Māmaki is the favored food of the caterpillar of the Kamehameha butterfly, *Vanessa tameamea,* the black and orange butterfly that is one of only two butterflies that are native to Hawai'i. The Hawaiians used the fibrous inner park of this plant to make one kind of kapa, and the mucilage from the bark was used to help glue layers of kapa together. The fruit was given to pregnant women and young infants to prevent thrush in the child, and also used to dress wounds. A tea was made from the leaves and drunk as a tonic. Health food stores still stock dried māmaki leaves for making tea.

Look for: A large shrub with large alternate toothed broadly oval pointed leaves that are green above and pale or reddish-veined beneath.

MANONO
Hedyotis terminalis
Rubiaceae (gardenia, coffee, pilo)

There are 22 species in this genus in Hawai'i, 20 of them endemic and two introduced aliens. I will describe 2 of the commoner native species, the ones I am most familiar with, in the hope that once you can recognize these, the others will be identifiable also. The species above is one of the most variable plants in the Hawaiian flora—probably only the 'ōhi'a exhibits so many different forms. The plant is a shrub, sometimes sprawling and almost a vine, and sometimes a small tree. The leaves are opposite pointed ovals about 1 to 7 inches long by ½ to 3 inches wide. When I was first learning to recognize our native plants, I was told that one distinguishing characteristic of the "rubes" (Rubiaceae) was that the new leaves emerging from the end of a stem looked like a pair of "kissing lips". This is particularly true in this genus, and something to watch

Look for: A shrub, viny shrub, or small tree with opposite leaves that look like "kissing lips" when first emerging, and small grappling hook flowers.

93

for. Numerous flowers appear in clusters at the tips of a branch. Each is a tube with 4 greenish-purple narrow petals sprouting from its end at abrupt right angles to it, or even folding back towards its base. A friend of mine says they look like miniature 4-pronged grappling hooks, and this is an apt description. This flower is about ¼ inch across. Clusters of small dark purple to black berries develop after flowering. This plant is found in mesic to wet forests on all the larger Islands.

The other species is *H. schlechtendahliana*, or kopa in Hawaiian. The Hawaiians gave a number of names to different species in this genus. Besides manono, which was used for several different species, they were called au, pilo, ʻāwiwi, kioʻele, ʻuiwi, and kamapuaʻa. This seems strange since we know of no uses that they had for these plants. Local names from different districts or Islands may be partly responsible. Kopa is a highly variable, small, slender, sprawling shrub or herb with nearly stalkless pointed oval leaves that are often almost heart-shaped. The paired opposite leaves are usually well separated along the slender stem. The flowers emerge from the tip of the stem and are pale green to greenish white or cream in color. The flower tube is ⅓ to ⅔ of an inch long with

KEN SUZUKI

Look for: A slender, sprawling shrub or herb with opposite leaves, often clasping the stem and small greenish to cream 4-petaled flowers, similar to those of manono.

4 narrow petals sprouting from it, much as in the manono above. This plant is found in mesic to wet forests on Kauaʻi and Oʻahu, and rarely on Molokaʻi, Lānaʻi, and Maui.

NAUPAKA
Scaevola spp.
Goodeniaceae (no familiar relatives)

There are over 100 species in this genus, most found in Australia. Eight endemic species and 1 indigenous one occur in Hawaiʻi. All except 1 are readily recognized by the very distinctive flower, which is usually present. The different species often exhibit interesting features, so I will mention the 6 that you are most likely to encounter.

Naupaka kahakai (beach naupaka), *S. (sericea) taccada*, is familiar to anyone who has ever strolled along the beach in Hawaiʻi. It is a fairly large, dense shrub with large, 6-inch long, oval glossy leaves and white flowers about ¾ of an inch across. The flowers look like they have been sliced in two—instead of a full circle of petals, the blossom is only a half circle. This is the typical form of our naupaka. The flowers are followed by white berries that float in sea water and will germinate after a year at sea. Thus it is not surprising that this species is indigenous and occurs in south east Asia and throughout the Pacific as well as in Hawai'i. The berries were occasionally eaten in times of famine, but I believe that you would need to be very, very hungry to make a meal of them. This plant was also sometimes called "huahekili" meaning "hailstones" because of the resemblance of the berries to hail.

Naupaka kuahiwi (mountain naupaka) was the name given to 4 different species. *S. gaudichaudiana* is a shrub or small tree found in mesic and wet forests on Kaua'i and O'ahu. It has fragrant white half-flowers which look just like those of the beach species. The leaves on this species are spindle-shaped, 2 to 4 inches long, light green, and serrated. *S. chamissoniana* is a similar species in wet forests on Moloka'i, Maui, Lāna'i, and the Big Island. All the mountain naupaka have purple-black berries, although occasionally the berries on S. gaudichaudiana will be white. *S. mollis* is usually found at somewhat higher wetter elevations than *S. gaudichaudiana*. Its leaves are often grayish in color and hairy underneath. The flower has a purple throat. It is found on Kaua'i and O'ahu. This species and *S. gaudichaudiana* seem to hybridize readily and produce fertile offspring, however, so you may see plants that are intermediate between them.

> Look for: A coastal shrub with large glossy oval leaves, white half-flowers, and white berries.

In contrast to the beach naupaka, the black berries of the mountain species are not edible under any circumstances. A friend of mine, Kost Pankiwskyj, a serious and very knowledgeable student of our native plants, likes to taste their fruits. He tells me that these berries are the most bitter, most awful tasting things he has ever sampled.

The remaining naupaka kuahiwi is a low shrub of the very dry forest. *S. gaudichaudii* (yes, it was named after the same botanist as *S. gaudichaudiana*) is found on dry ridges and flats on all the main Islands. It has small, toothed, stiff leaves 1 or 2 inches long and brownish-yellow half-flowers with narrow sharply pointed petals, and purple-black

JOHN HOOVER

> Look for: A shrub or small tree of mesic or wet forests with white or purple half-flowers and black berries.

berries. On Oʻahu, this plant is uncommon and found in only two locations that I know of. You are most likely to see it along some of the leeward jeep roads above Kaʻena Point, near the Kuaokalā Trail.

ʻOhe naupaka, *S. glabra*, is a shrub of very wet forests on Kauaʻi and Oʻahu. Its flowers are a beautiful bright yellow and so long and curved that it was initially mistaken for one of the lobelia family. The half-flower character of these blossoms is not very evident. Like many of our native lobelias and mints, it was probably adapted to pollination by honey creepers, birds with long, curved bills. This lovely shrub is most easily seen along the boardwalk trail to Kilohana Lookout on Kauaʻi.

The unusual half-flowers of naupaka excited the imagination of the Hawaiians and a number of legends were devised to explain them. In one, two lovers quarreled, and the woman in a fury tore what until then had been an intact, circular naupaka flower in two. She told her lover that she would never forgive him until he could

Look for: A small, dry-land shrub with yellowish half-flowers that have narrow, sharp petals and black berries.

find a whole flower again. But the gods had changed all the naupaka flowers of beach and mountain into half-flowers and he could never find a whole one, and eventually died of a broken heart. Another legend claims that a lovely stranger became infatuated with a village youth. He dallied with her awhile, but then went back to his former girlfriend. The stranger pursued them with fiery eyes and tore them apart, and they realized that they had offended the volcano goddess, Pele! She chased the youth into the mountains on a wave of molten lava, but the gods took pity on him and turned him into the naupaka kuahiwi. (I've never understood how this saved him. Perhaps Pele didn't feel she'd get much satisfaction from incinerating a mere plant!) In a rage, Pele turned and pursued the lass down toward the shore. The gods turned

Look for: A shrub of very wet forests with striking curved yellow blossoms.

the maid into naupaka kahakai. And ever since then, the lovers have been parted, each with half a flower, yearning for the other, one on the mountain, the other on the shore, never to be united again. (Martha Beckwith says that a Hawaiian legend involving romance that has a happy ending has probably been edited by Westerners. The Hawaiians didn't *believe* in happy endings.)

ʻŌHĀ WAI
Clermontia spp.
Campanulaceae (lobelia, hāhā, campanula)

This family has 6 endemic genera in Hawaiʻi, plus *Lobelia* itself which is in-digenous. There were over 100 species of these plants, and the radiation of all of these plants from perhaps only 5 original colonists with most of them descending from just 1 of these, is one of the marvels of island evolution. When the first Europeans arrived in Hawaiʻi, these plants were still widespread and numerous, but the effects of herbivorous mammals, insects, and molluscs, the loss of the birds that once pollinated them, and competition from introduced weeds have taken a sad toll and now many of these lovely natives are extinct or very rare. I will only consider 2 groups in this book: the shrubby clermontias which I discuss here, and the cyaneas and others that resemble them, all of which look like miniature palm trees (see hāhā).

There are 22 species of *Clermontia* listed in the *Manual*, 1 of which is extinct, and 4 very rare. Most species are found on Hawaiʻi or Maui, with fewer on the older Islands—Kauaʻi has but one. These are branching woody plants growing as shrubs or small trees, with alternate spindle-shaped leaves, 6 to 10 inches long, sometimes with toothed edges. Leaves tend to cluster near the branch ends, and prominent round leaf scars in a closely packed, regular pattern, are left where old leaves have fallen off. The branches tend to curve upwards, like the arms of a candelabra, and the young branches are often thick, soft fleshy and easily broken. Two or 3 flowers branch from a common stalk and are an inch or 2 long, curved like the bills of the nectar-feeding birds that once pollinated them, and split down the back. There are usually 5 or 10 petals to each flower and they range in color from green to white, rose, purple, or even a purple so dark that it is almost black. The fruit is about the size of a quarter, yellow or orange, and often slightly ribbed so that it looks like a tiny pumpkin. This is the member of the lobelia family that you are most likely to see and to be able to recognize on

KEN SUZUKI

Look for: A branching shrub with spindle-shaped leaves clustered near the branch ends, which are thick and fleshy. Prominent leaf scars pattern the branches, especially near the tip. Flowers are fairly large, curved, and split down the back, with 5 or 10 petals.

our trails. ʻŌhā wai is found in upper mesic to wet forests on all the higher islands.

The Hawaiians ate the fruit of some species of this plant and cooked the young leaves of others for greens. The sticky sap of one species was used to trap native birds to secure their feathers for the feather-work cloaks, helmets, and idols of the ali'i.

PILO
Coprosma spp.
Rubiaceae (coffee, cinchona - quinine, noni, nānū - gardenia)

To the expert botanist, the pilo are all united in one genus by the characteristics of flower and fruit. It is much more difficult for the amateur to find striking features that unite the different species in a common group. Indeed, several of them look so different that the Hawaiians gave them entirely different names. Once you have become familiar with the plants, you will begin to see what they have in common, but there is no simple test, or one that applies readily to all of them. For this reason, I will attempt to explain what I think they have in common, and then consider the species that you are most likely to encounter individually and point out the distinguishing features of each. There are 13 endemic species in this genus in Hawai'i, plus one indigenous one. You are likely to see 5 species, and these are the ones I'll discuss. Most pilo are viny shrubs with many branches—in most, young side branches seem to begin to grow from the stem at the base of nearly every pair of opposite leaves. Most have orange berries when in fruit, but male and female flowers are usually on separate plants, so not all will have berries.

Look for: A viny shrub with light green opposite leaves, orange berries, and new branches emerging from nearly every leaf base.

C. foliosa, pilo, is a sprawling shrub of mesic to wet forest on all the islands except Hawai'i. It looks much like maile, but has well-spaced, lighter green, less glossy leaves that are pointed ovals about 1 to 3 inches long, orange berries, and the tendency to branch at every leaf node.

Look for: The whorl of 3 untoothed leaves emerging from the growing tip of the stem, and orange berries.

C. longifolia (not given a Hawaiian name in the *Manual*) is found in mesic to wet forest only on O'ahu. It has orange berries, but is different in almost every other way from most pilo. The spindle-shaped leaves emerge from the growing tip of the branch in a whorl of 3, and the mature leaves on the stalk are also arranged in threes around the same point on the stem. Leaves are 2 to 5 inches long and an inch wide, and, like those of all pilo, lack teeth. The shrub does not branch frequently like other pilo. However, the whorl of 3 leaves emerging from the end of the branch is so distinctive, and so different from any other plant you are likely to see, that there is little risk of mistaking this plant.

C. montana (no Hawaiian name given) is a small tree or large shrub of the sub-alpine shrubland of Haleakalā or the Big Island, often forming a dominant component of this shrubland. The leaves of this plant are small, leathery, bluntly pointed ovals on short lateral branches. It has the typical orange berries and tendency to branch profusely that characterize the pilo.

Look for: A common, densely branched shrub of sub-alpine country with orange berries and the typical pilo branching habit.

C. ernodeoides—kūkaenēnē (nēnē dung) is a creeping, low-growing plant of sub-alpine lava and cinder fields on Haleakalā and the Big Island. It has small, dark green, narrow, closely spaced leaves less than ½ inch long that are clustered along short ascending side branches. Its most striking, and unmistakable feature, is the presence of numerous glossy black berries on the female plants. These are eaten by the native goose, the nēnē.

JOHN HOOVER

Look for: A low growing, creeping plant with small dark green leaves and prominent glossy black berries.

C. granadensis, mākole. This plant is different enough that until very recently it was placed in its own genus, *Nertera*. I understand that it is now considered a *Coprosma*, however. It is the only indigenous species considered here, being found in Columbia, Indonesia, and on the Juan Fernandez Islands as well as in Hawai'i. It is a lovely light green, spreading creeping herb found on mossy banks and tree trunks in very wet forests on all of the larger Islands (Mt. Ka'ala on O'ahu). It has many branches and orange berries. The leaves are small, opposite, and oval.

Look for: A ground-hugging, light green, creeping herb with orange berries on mossy terrain in very wet forest.

99

Lorin Gill points out that the Hawaiian word "pilo" means a bad odor—not skunk bad, but unpleasant like an arm pit or stagnant water. It is interesting that the scientific name for this genus has a similar connotation—"kopros" means dung, and "osme" is smell, and was given to plants in this genus because many of them have a fetid odor when crushed. Therefore, it is curious that the Hawaiian pilo do not have a bad smell. Apparently, the ancestors of the Hawaiians were acquainted with plants in this genus in their original homeland in Southeast Asia or the nearby islands, that did have a bad odor, and retained their name for such plants as they encountered related species in their migrations through the Pacific until they reached Hawai'i!

Thomas H. Rau

Look for: A many-stemmed shrub with inch wide, 5 petalled, rose pink flowers, purple berries, and opposite leathery leaves with 3 main veins.

ROSE MYRTLE
Rhodomyrtus tomentosa
Myrtaceae ('ohi'a, guava, eucalyptus)

Rose myrtle is a shrub up to 10 feet tall. It has blunt oval opposite leaves 1 to 4 inches long that are fuzzy beneath. Each leathery leaf has 3 prominent veins running from stem toward the tip, similar to the pattern seen in the melastome family, which includes *Clidemia* and *Miconia*. Numerous flowers each about 1 inch across have 5 petals and are rose pink in color. They are followed by edible purple berries ¾ inch long. This plant is native to India and southeast Asia. It was reportedly introduced to Kaua'i before 1920, and is now spreading and threatening to become a serious pest in mesic to wet forest on all the major Islands.

KEN SUZUKI

Look for: A dense shrub with narrow needle-like leaves and numerous white or pink 5 petalled flowers that are about ½ inch across.

TEA TREE, MANUKA
Leptospermum spp.
Myrtaceae (guava, myrtle, eucalyptus)

There are 3 species in this genus in Hawai'i. They are shrubs or small trees with narrow, needle-like leaves and ½ inch wide 5-petalled white or pink flowers. The fruit is a small dry capsule which opens at the top along lines in a star pattern. These plants are native to Australia and New Zealand and were introduced to Hawai'i about 1927 as ornamentals. They are now spreading and threaten to become major invasive weeds in mesic and wet forest on Kaua'i, O'ahu, and Lāna'i.

The early European settlers in Australia and New Zealand apparently made a tea from the aromatic leaves of these plants, thus giving them the common name. The vegetative growth of these plants resembles that of the native pūkiawe. They can be told apart, however, by the following traits: (1) Tea tree blossoms are usually present and quite conspicuous while those of pūkiawe are very small and difficult to find. (2) The pūkiawe fruit is an attractive white to red berry while that of the tea tree is a dry capsule with a star-shaped opening on top.

THIMBLEBERRY and ĀKALA
Rubus rosifolius, R. hawaiensis
Rosaceae (rose, raspberry, 'ūlei, strawberry)

Seven species in this genus are found in Hawai'i, 2 endemic natives and 5 introduced weeds. You are most likely to encounter the first species above, which is a low sprawling, rather delicate viny shrub covered with slender sharp prickles. As the species name implies, the alternate leaves are like those of a rose, being odd-pinnately compound, usually with 2 or 3 pairs of sharply pointed oval leaflets along the stem and a single, slightly larger one at the tip. Each leaflet is 1 to 3 inches long and bears numerous small sharp teeth along each edge. White 5-petalled flowers about 1 inch across are followed by bright red berries like those of a raspberry and about an inch, or a bit less, long. These are sweet and edible, tasting like a mildly flavored raspberry and full of tiny seeds. The plant originally came from Asia and is now widely spread, sometimes in dense patches, in mesic to wet forest in Hawai'i. It is thought to have been introduced

JOHN HOOVER

Look for: A low, prickly, sprawling shrub with rose-like, alternate, odd-pinnately compound leaves, white 5-petalled flowers, and raspberry-like fruit.

Look for: An erect, sometimes sprawling plant with compound leaves with 3 toothy leaflets, dark pink flowers, and large raspberry-like fruits.

from Jamaica in the 1880s. This plant is spread by birds and other animals and is a moderately serious pest in our forests, obstructing trails with its prickly canes and crowding out desirable vegetation.

R. hawaiensis, or ʻākala, is a fairly common endemic raspberry that grows in high elevation moist forest on all the larger Islands except Oʻahu. It is a much larger plant than thimbleberry, with sturdy, usually erect, sparingly branched canes up to 15 feet tall. The leaf may be 6 inches long, but there are only 3 leaflets in this species, and the plant may or may not have prickles. As in thimbleberry, the leaflets have many teeth along each edge. ʻĀkala flowers are an inch or more across and a dark pink. The fruit is an inch or two in diameter, red or purple, or rarely yellow, and tart. It is edible, but not very flavorful to my taste. A good place to see this plant is behind Palikū cabin in Haleakalā National Park, but it is not uncommon and will be encountered fairly frequently in mesic to wet forests on the outer Islands. The Hawaiians ate the fruit when they found it, but since the plant grew far from their usual population centers, it was not a common treat. They also obtained a rose-colored dye for kapa cloth from the fruit. ("Kala" means "pink".)

The other endemic species is uncommon and similar to ʻākala, while several of the remaining introduced aliens, various kinds of blackberries or raspberries, are very serious invasive pests, forming dense thickets of thorny canes, spreading, and crowding out native plants.

TI, or KĪ
Cordyline fruticosa
Agavaceae (hala pepe, agave, New Zealand flax)

KEN SUZUKI

Few residents of Hawaiʻi who have lived here for more than a month or two, can be unfamiliar with the ti plant, as it is widely grown in people's yards and used to landscape public buildings, hotels, and offices. This plant was introduced to the Islands by early Polynesian settlers, and is found widely spread in shaded areas in the mesic forests in our mountains. The plant has a long, spindly sparsely branching stalk topped by a whorl of long oval leaves, 18 to 30 inches in length and 3 to 7 inches wide. Occasionally, sprays of pinkish-white 6 petalled flowers emerge

Look for: A slender, little-branched plant with a tuft of long oval leaves in a spiral at the tip of each branch, like a feather duster.

from the center of the leaves. Fruit is rarely seen in Hawaiʻi, and the plant probably propagates itself by vegetative spread. In wild colonies, sprawling stalks often have thick, knobby roots reaching toward the ground on their lower sides, and these may help to establish new plants once they make contact with the soil.

The large, flexible, hairless and odorless leaves were very useful to the Hawaiians for wrapping food and all kinds of small parcels. They were used to line the imu for cooking and were second only to pili grass as thatch for the houses. They could be twisted into lei or for ropes, and the cables were coiled to make sandals when rough lava fields had to be crossed. It was believed that the plant would ward off evil spirits and it was planted around most houses for this purpose, while lei and wristlets made of ti could be worn for protection. The leaves were tied to nets to make rain capes. Children used to break the top off a ti plant, sit on the tuft of leaves, hold the stem, and slide down a steep, muddy bank or grass slope as a form of sledding.

Various parts of the plant were mixed with other herbs to treat insomnia, kidney problems, asthma, sinus congestion, tuberculosis, venereal disease, vaginal discharges, and as a wrapping for poultices. Feverish patients could be cooled by having ti leaves placed on their bodies. The roots, which can grow to be very large, are rich in fructose and after long steaming in the imu produce a sweet, molasses-like treat. They were sometimes cooked in this way as a famine food, when other crops were in short supply. After Western contact, this sugary preparation was diluted, allowed to ferment, and the resulting alcohol distilled using iron blubber-rendering pots from whaling ships to produce a potent liquor called "ōkolehao", "ōkole" meaning "bottom" and "hao"," iron", referring to these pots.

TREE DAISY
Montanoa hibiscifolia
Asteraceae (daisy, sunflower, lettuce, thistle)

KEN SUZUKI

This shrub is a pithy-stemmed plant up to 12 feet or more in height. The opposite leaves can take a number of forms, but are often deeply lobed with irregular, large teeth and sharply pointed tips. They may be from 3 to 20 inches long by 1 to 14 inches wide, and have soft hairs on both surfaces.

Look for: A large shrub with irregularly lobed, somewhat hairy leaves, and clusters of 1 inch wide daisy-like flowers which bloom in midwinter.

The daisy-like flowers, about 1 inch in diameter, are borne on branching stalks and have 6 to 10 white or rose-tinged petals with a yellow center. This shrub is cultivated as an ornamental for its flowers, which appear in midwinter. It was introduced into Hawai'i in 1919 by E. M. Ehrhorn, and has since become sporadically established in the wild in dry to mesic forest on all the larger Islands. It appears to be spreading. It is native to Central America.

TREES

AFRICAN TULIP TREE
Spathodea campanulata
Bignoniaceae (catalpa, jacaranda, pink tecoma, gold tree)

These are medium sized trees up to about 70 feet tall with a light brown trunk and opposite odd-pinnate compound leaves. There are usually 1 to 8 pairs of leaflets, with the odd one at the tip, each being a pointed oval 3 or 4 inches long by 1 ½ to 2 inches wide. The leaflet edges may turn down slightly. The leaflets have sunken veins and there are rust-colored hairs on the lower surface. Large showy irregularly shaped red-orange (rarely yellow) blossoms 4 inches wide are followed by canoe-shaped pods 6 to 9 inches long that occur in erect clusters. The tree may be in bloom

Look for: Medium sized trees with odd pinnate compound leaves and conspicuous showy orange-red blossoms followed by upright clusters of finger-like pods.

throughout the year. This plant is native to tropical Africa. It was first introduced into Hawai'i as an ornamental by Hillebrand in 1871, but the trees did not produce viable seeds and Joseph F. Rock reintroduced it from Java in 1915. Since then it has been used on a modest scale for reforestation, with about 30,000 trees being set out on all the main Islands. It is now found scattered through our forests in mesic and wet sites, and is spreading, though it is not yet clear if it is likely to become a problem.

Toona and tropical ash are common trees with similar compound leaves, but these differ from those of the tree above in lacking the sunken veins, rusty hairs, and turned-down leaflet edges.

'AHAKEA
Bobea spp.
Rubiaceae (coffee, gardenia, ixora, pilo, manono)

This genus is endemic to Hawai'i and contains 4 species. These are trees 30 to 40 feet tall with pointed oval opposite leaves 2 to 6 inches long and 1 to 2 inches wide. The leaf has a prominent yellow midrib and obvious lateral veins. The tree somewhat resembles the related kopiko, but the leaves are lighter colored and more translucent, and the young leaves in particular have a yellowish tint. Emerging leaves form a pair of "kissing lips" that is typical of a number of the "rubes". The leaves tend to cluster near the branch tips. The trees are found in mesic to wet forests. One to 3 small greenish to white tubular flowers emerge on stalks from the base of the leaves. These are followed by small oval purplish-black juicy fruit. Flowers and fruit are only seen occasionally on these trees, however.

Kost Pankiwskyj tells me that the fruit of 'ahakea is very bitter and ranks second only to that of the mountain naupaka for obnoxious flavor. This may be a diagnostic characteristic for the tree, but I don't recommend using it.

Bark of this tree was mashed and mixed with water to form a preliminary wash in the treatment of abscesses and infected wounds. It was also part of a complex mix taken to "clean the blood". The wood is

Look for: A tree of mesic and wet forests with leaves that are somewhat yellowish, especially when young.

very hard and durable and so was highly preferred for door frames and doors and especially for the gunwales of canoes, which were subject to much abrasion. Its attractive yellow color which turns from a dull orange brown to a dark gold when rubbed with kukui oil made it a standard trim for canoes of any importance, and even in modern days, the gunwale of the Hawaiian canoe is colored yellow to continue this tradition.

'ĀLA'A, ĀULU
Pouteria sandwicensis
Sapotaceae (sapodilla, sapote, egg fruit)

Look for: A tree of lowland gulches with blunt-tipped stiff dark green glossy leaves and milky sap.

This endemic plant is a tree 40 feet tall or more with milky sap and leaves arranged in a spiral around the twig. The leaves are fairly stiff oblongs 2 to 6 inches long or more, and blunt or indented at the tip. They are a glossy dark green on top and may have a fine coat of rust-colored hairs beneath, especially when young. The globular fruit may be yellow, orange or purple and is 1 to 1 ½ inch in diameter. The trees are found in mesic forest, especially in lowland gulches, on all the larger Islands.

The seeds of this plant were used in lei. The milky sap was occasionally used as a bird lime to trap the small forest birds whose feathers adorned the spectacular capes of the ali'i and other artifacts. Wood from 'āla'a was used in house construction, for 'ō'ō or digging sticks, the main agricultural implements of the Hawaiians, and for spears.

ALLSPICE
Pimenta dioica
Myrtaceae ('ōhi'a, eucalyptus, myrtle)

This tree has opposite glossy green oblong leaves 4 to 6 inches long by an inch wide. These are rather brittle and when crushed emit a powerful fragrance of allspice, which is probably the most reliable way to identify the plant. The trunk has smooth, light brown mottled bark, rather similar to strawberry guava but paler and perhaps a little more yellowish in color. White, fragrant flowers are followed by clusters of small round fruit. Both flower and fruit are about ¼ inch in diameter. The spice is obtained from unripe, dried berries. The plant originally comes from Central America and was introduced into Hawai'i from Jamaica in 1885. It is now naturalized in mesic forests on O'ahu and Kaua'i.

KEN SUZUKI

Allspice is one of only 3 major spices that originated in the Americas and are now widely used in international cuisines. The other two are chili peppers and vanilla, which is obtained from the fruit of an American orchid. A close relative of allspice, the bay rum tree, *P. racemosa*, has somewhat broader leaves with edges that roll under slightly and have a distinctive scent when crushed. This tree is from the islands of the Caribbean and was also introduced into Hawai'i in 1885 from Jamaica. It is now naturalized and spreading in Moanalua Valley on O'ahu.

Look for: A tree with opposite glossy green brittle oblong leaves that have a strong odor of allspice when crushed.

ARDISIA and HILO HOLLY
Ardisia spp.
Myrsinaceae (kōlea)

There are two introduced ardisias in Hawai'i, but they look so different from each other that it is necessary to describe them separately. *Ardisia elliptica* is sometimes called "shoebutton ardisia" but most people just call it ardisia, and

I will follow this practice. This weedy invader is a tree of moist mesic areas, originally found in sheltered valleys, but now spreading onto the ridges. It came from Sri Lanka and may have been introduced by Hillebrand in the mid 1800s. It is an erect, straight small tree, or shrub, up to 10 or 15 feet tall, with slender branches that are perpendicular to the trunk. The alternate leaves are elliptical, pointed, and about 4 inches long. Young

Look for: A small, straight tree with slender perpendicular branches and colorful young leaves bearing multicolored clusters of berries.

leaves are a bright pink in color. Flowers occur in clusters and are star-shaped, pinkish-white with 5 petals and about ½ inch across. The clumps of ¼ inch berries are often a mixture of red (immature) and black (ripe) fruit. This plant has become an increasingly common pest in recent years.

Ardisia crenata or Hilo holly is a small, usually unbranched plant, about 2 or 3 feet tall. It is found scattered in mesic to wet forest. The plant is native to south Asia and was cultivated in the Kamehameha Schools nursery about 1930. The dark green, crenellated leaves in a bunch at the top of the plant are somewhat reminiscent of English holly, but it is probably the clusters of bright red holly-like berries that gave it the name "Hilo holly". The leaves do not have the spiny edges of English holly, however.

Look for: A short plant with a dense dark green tuft of crenellated leaves at the top and clusters of striking red berries.

AVOCADO, ALLIGATOR PEAR
Persea americana
Lauraceae (cinnamon, camphor, bay, sassafras)

Look for: A moderate sized tree of mesic forests with large, alternate, pointed oval leaves and large pear-shaped or round fruit.

This common fruit tree is frequently seen along our trails in low elevation mesic forests. It has large alternate pointed oval leaves, 4 to 10 inches long by 2 to 4 inches wide. The tree is most easily recognized when in fruit by the large pear-shaped or round avocados hanging from the branches. The plant is a native of Central America and now

107

occurs in mesic forest on all the larger Islands. It was probably introduced to Hawai'i early in the 1800s by Don Francisco de Paula Marin, and is known to have been part of the cargo of plants brought to the Islands from Rio de Janeiro on the HMS Blonde in 1825.

BAMBOO, 'OHE
Poaceae (grasses, sugar cane, and all the major food grains)

The *Manual* lists 2 species of bamboo that are naturalized in Hawai'i, the black bamboo, *Phyllostachys nigra*, which appears to be the more common and widespread, and 'ohe, *Schizostachyum glaucifolium*, which was probably introduced by the Polynesians, although it might be indigenous. Black bamboo is native to China and was introduced as an ornamental. It spreads by means of underground stems (rhizomes) and has formed extensive stands on moist, shaded slopes and stream banks. It was first collected here in about 1951. Despite the name, the species in Hawai'i is green when young, turning orange or yellow as it ages. The tree is 30 feet tall or more and 2 to 4 inches in diameter. One variety has yellow canes with vertical green stripes.

'Ohe is very similar in appearance, forming open clumps. The canes have long spaces between nodes and are thin-walled. This bamboo occurs in many of the islands of Polynesia south of Hawai'i. It is found in shaded sites along streams in mesic valleys. Many bamboos look very similar and it is difficult to distinguish different species without the flowers, which are seldom seen.

Extensive bamboo thickets have formed on all the Islands. While they rarely seed, bamboo slowly spreads as the underground shoots push out from the edges of a clump and send up new canes. The stands are dense and dark so that almost nothing else can grow in a bamboo grove, and it is tedious to try to walk through one. They are very difficult to eradicate, and if unchecked may eventually displace all other vegetation, native or otherwise, in favorable habitats. This being said, walking along a narrow trail through the twilight of a dense bamboo grove, or scrambling through the jackstraw tangle of fallen canes off the trail, with the wind sighing through the tops and the canes swaying and clacking in the breeze, is an unforgettable experience. Ac-

Look for: It is hard to believe that anyone does not know what bamboo looks like, but if you do not, look for a giant grass with tall straight smooth canes that have nodes at regular intervals.

cording to some botanists, all bamboo of the same species, all over the world, flower and go to seed at the same time and then die. If this is true of our species, it may be possible to reclaim some of the territory that they now preempt once this happens, but in the 45+ years I have lived here, it hasn't happened yet! However, some species are said to flower only once in every 120 years or so, so perhaps it may in the future.

People harvest the young shoots of our commonest bamboo for food and use the canes as walking sticks and many other things where a light straight strong pole is useful. The Hawaiians believed that ʻohe was a kino lau, or plant form, of Kāne. They used the bamboo for styluses for drawing designs on kapa cloth or as stamps for creating other patterns on it. Kukui nuts were placed in a hollow cane to make a lamp, and several different kinds of musical instruments—a rattle, stamping tube, and nose flute—were made of bamboo. The ashes of the leaves were used in a concoction to produce a poultice for cuts, bruises, and sores. The sharp edge of a split cane could be used as a knife. It is said that the town of Kāneʻohe (bamboo man) on Oʻahu got its name from an incident in which a woman living there was asked about her abusive husband, and replied that he was "a man as sharp as a knife".

BLACK WATTLE
Acacia mearnsii
Fabaceae (koa, indigo, gorse, peanut)

This is an Australian tree that has been used for reforestation in Hawaiʻi. The Division of Forestry planted about 65,000 of these trees in the Islands, primarily at Kula, Maui, and Mokulēʻia, Oʻahu. Unfortunately, they have now spread and become serious pests. This tree may exceed 100 feet in height and can be readily identified by its doubly compound leaves, in which 8 to 21 pairs of stemlets branch from the main leaf stem and each bears 30 or more pairs of tiny leaflets. The leaves are alternate and gray-green or dark green in color. The flowers are small light-yellow fuzz balls about ¼ inch in diameter and are followed by clustered pods 2 to 4 inches long and ¼ inch wide, flattened slightly between seeds. The first trees were apparently brought in from a San Francisco nursery in

Look for: Trees with gray-green, doubly compound leaves with tiny (pencil-line wide) leaflets.

1911, and are now found infesting pastures and dry to mesic forests on all the major Islands.

Like many other legumes, this tree forms nitrogen-fixing associations with bacteria, thus enriching the soil beyond what the native plants require and encouraging the invasion of weeds. It responds to fires or cutting of the tree by sending up sprouts from the stump and roots and can thus form dense thickets that are a problem in pastures and crowd out other species in the forest. It also produces enormous numbers of seeds which aid in its spread.

BRUSH BOX
Lophostemon confertus
Myrtaceae (eucalyptus, 'ōhi'a, guava, mountain apple)

These trees, closely related to eucalyptus, are up to 60 feet tall with straight trunks 1½ feet in diameter or more, and narrow rounded crowns with dense foliage. The leaves are alternate, but tend to be clustered at the ends of branches where the most recent leaves often form a whorl of 4 or 5 around the end of the twig. The lance-shaped leaf is 3 to 8 inches long by 1½ to 2 inches wide, dull green in color with light yellow mid rib and very fine side veins. Three to 7

Look for: A straight sizeable tree with 4 to 5 lance-shaped leaves, 3 to 8 inches long, attached in a circle around the end of the twig, and cup-shaped woody capsules about 1/3 inch across.

flowers cluster near the end of a short, flattened, unbranched stalk. They are white and fragrant, with 5 petals and about an inch across. The fruit is a woody, cup-shaped capsule about ⅓ inch in diameter. Brush box is native to Australia. Since 1929, it has been extensively used in forestry plantings in Hawai'i, with over 400,000 being set out on the larger Islands. It is now naturalized and spreading in the wild, at least on O'ahu.

CAIMITILLO, SATINLEAF
Chrysophyllum oliviforme
Sapotaceae (āla'a, sapodilla, egg fruit)

These introduced trees are 20 feet tall or more, slender, and with oblong pointed leaves about 2 to 5 inches long by 1 ½ to 2 ½ inches wide. The leaves are alternate and dark green above but covered with a rust-colored fuzz underneath. Small tubular 5-petalled flowers may cluster around the stem at the base of the leaf, and are followed by oval purple-black fruit about 1 inch long. These are not edible. The tree is native to the Bahamas, the southern tip of Florida, and some Caribbean islands. It is often planted as a street tree in Hawai'i and has become established in mesic to moist forest on the Islands.

Look for: A slender tree with alternate leaves that are dark green above and covered with rust-colored fuzz beneath.

CECROPIA, TRUMPET TREE
Cecropia obtusifolia
Cecropiaceae (no familiar relatives)

These are very open, scraggly trees about 15 to 35 feet tall with round leaves about 8 to12 inches across that are deeply divided into 9 to 15 lobes. The lobes broaden toward the tips. The leaves are found in bunches at the branch tips. The lower surface of the leaf is hairy and nearly white in color. The tree has soft wood and the stem and branches have hollow chambers. In its native lands, some species house ants in these chambers and feed them with small fat- and starch-rich bodies that form at the base of the leaf stem. The ants pay for the board and room by keeping the tree free of herbivorous insects. The tree is fast growing and short lived. The sparse canopy casts little shade. Cecropia species are native from southern Mexico to Ecuador. This tree was first noticed in Hawai'i about 1926 and is now naturalized and spreading in mesic and wet forests on most of the larger Islands, where it threatens to become a pest.

Look for: A very open, slender, leggy tree with round, deeply lobed leaves clustered at the branch ends. The leaves are white beneath.

CHINESE BANYAN
Ficus microcarpa
Moraceae (wauke, mulberry, figs, breadfruit, marijuana)

More than 30 different ban-yan species have been planted in Hawai'i, but only 3 or 4 set viable seeds and are able to spread. This is the most ubiq-uitous of those that do, being able to sprout in any crack in a wall, in a rain gutter, on other trees, or anywhere else that the tiny seeds can lodge. It is a real nuisance around Honolulu and other towns, and is also making its ap-

Look for: A gray-barked tree with aerial roots, milky sap, and a dense, spreading crown.

pearance in our forests where it promises to be an increasing problem. The tree has smooth light gray bark, milky sap, and a dense, spreading crown with small pointed oval leaves that are 2 inches or more in length and about an inch wide. They are alternate and emerge from a small, conical, sharply pointed tip that is covered with a sheath. The leaves are often disfigured by a gall wasp, *Josephiella microcarpae*, whose larvae feed on them, but apparently do little damage to the trees. A pair of small round yellow or red fruits about ¼ inch in diameter hug the leaf bases in season and are readily consumed by birds who spread the seeds. Chinese banyans send down thin filamen-tous aerial roots which will develop into auxiliary trunks once they reach the ground. The tree is native to south Asia, Australia, and adjacent islands and has been grown in Hawai'i since the early 1900s.

Like other figs, this plant can not set seeds until the flowers are pollinated by a particular species of tiny wasp, which explains why most of our banyans are unable to spread, and it has become naturalized only since 1938 when the proper wasp became established here, due to the ill-advised efforts of Dr. Harold Lyon and the Hawaiian Sugar Planters Association. Now it is slowly and insidiously invading our dry and mesic forests where the dense shade, and perhaps allelopathic properties of its fallen leaves, prevent the growth of most other plants under it. (See the discussion of strawberry guava for an explanation of allelopathy.) This tree, along with the 2 or 3 other banyan species whose pollinating wasps have become established in Hawai'i, is now present in forests on all of our larger Islands. These are the trees that line the famous Banyan Drive in Hilo.

CINNAMON, PADANG CASSIA
Cinnamonum burmanni
Lauraceae (avocado, camphor, laurel, sassafras, dodder)

This introduced tree is common and spreading in wet and mesic forests in Nu'uanu, Tantalus, and Mānoa on O'ahu. It has glossy dark green pointed elliptical leaves with three prominent veins running from the stem toward the tip. New leaves are a coppery-red. Mature leaves are up to 4 inches long by 1 or 1½ inches wide. In mature trees the bark is black. The tree is a native of Indonesia and has been cultivated in Hawai'i for most of the 20th century although it was first collected in the wild here about 1975.

Look for: A tree of wet and mesic forests with glossy dark green elliptical leaves with 3 prominent veins running from the stem toward the tip, and black bark.

This tree, as Padang cassia, is cultivated on Sumatra where the bark from young twigs is harvested and sold as a spice, Korintji cinnamon, in the U. S. Another species, *C. verum*, from Sri Lanka, is the source of true cinnamon, however. *C. camphora*, from northern China, Taiwan, and Japan, yields camphor. These last 2 species are planted in Hawai'i, although not as commercial crops, as far as I'm aware, and are not likely to be seen along our trails.

COFFEE
Coffea arabica
Rubiaceae (ixora, gardenia, noni, pilo)

This plant was introduced to the Islands in 1813, probably by Don Francisco de Paula Marin, and by 1850 large amounts of coffee were being exported. Coffee has been a major crop in Hawai'i ever since. The tree has also gone wild and is commonly seen in mesic to wet valleys and shaded slopes in our moun-

Look for: A small erect tree with slender horizontal branches and 3 to 8 inch long glossy green leaves with prominent veins. Clusters of white flowers or red berries hug the branches in season.

tains, often forming dense, but not usually very extensive, stands. Coffee is a small tree about 15 feet tall with slender, horizontal branches bearing glossy dark green opposite leaves with prominent veins. The leaves are 3 to 8 inches long and 2 to 3 wide and are pointed ellipses in shape. The underside of the leaf is lighter than the top, and they have undulating edges. The bark is pale, nearly white, and rough to the touch. In season, white flowers cluster around the stem at the base of the leaves and soon produce red oblong berries about ½ inch long. The tree probably came from Ethiopia originally, but has long been grown throughout the tropical world.

COOK PINE
Araucaria columnaris
Araucariaceae (bunya-bunya, Norfolk Island pine, hoop pine)

This is a large straight symmetrical erect tree with a sturdy trunk and pointed spire that may grow to 200 feet tall. Its bark is gray, rough and thick. It peels from the trunk in horizontal hoops. Five, 6 or more branches emerge in a ring around the trunk at right angles to it and at regular intervals. Twigs sprout from each side of a branch in a single, horizontal, plane. In young trees, the leaves are tightly clustered, sharp, and needle-like. In mature trees the leaves are more blade-like, broadly attached, overlapping, arranged in spirals, and curved toward the twig, forming a sheath around it. Cook (*not* Cook Is-

land) pines have been used extensively for reforestation in Hawai'i and groves of these trees will often be encountered near the beginning of a ridge trail in mesic forest. I have not found any information on when they were introduced to Hawai'i or how many have been planted, however. The trees are native to New Caledonia where they were discovered by Captain Cook and were named after him.

Captain Cook was very pleased when he observed these trees since the British navy was always in need of tall straight masts for their sailing ships, and these pines seemed well suited to this purpose. He also discovered the closely related Norfolk Island pines, on the island for which they are named. These trees are so similar that it is difficult to tell them apart, but we now believe that most of the trees in Hawai'i

Look for: Tall, straight sharply pointed cone-shaped trees with gray rough bark, branches emerging at regular intervals in rings around the trunk, and tightly clustered scales or needles.

are Cook pines. They can easily be distinguished from the sugi by the difference in bark color, and the very regular pattern of branching in the araucarias. As in the case of the sugi, although these trees are often called "pines" they actually belong to a more ancient family than the true pines.

EUCALYPTUS, SWAMP MAHOGANY
Eucalyptus robusta
Myrtaceae (mountain apple, 'ōhi'a, guava, Java plum)

There are between 500 and 600 species of eucalyptus, almost all native to Australia, and more are still being described. More than 90 species have been used for forestry plantings in Hawai'i, but the *Manual* discusses only the 30 that seem to be replacing themselves or even spreading. Many of these trees are difficult to tell apart, and I will describe only the one species given above, on the assumption that once you can recognize this eucalypt, the others will be identifiable as members of this genus also. (The photo is not of this species, but is of a typical eucalypt.) Eucalyptus leaves often have a characteristic fragrance when crushed. See "paperbark", "brush box", and "turpentine tree" for other close relatives.

E. robusta, also known as "swamp mahogany", is a fairly large tree, growing to be 80 to 160 feet tall and 3 or 4 feet in diameter. It has rough, very thick, spongy fibrous bark that is deeply furrowed in mature trees. The bark is reddish brown in wet areas and grayish brown in dry. Mature leaves are alternate and broadly

Look for: A large straight tree with deeply furrowed spongy bark, red brown in wet areas and gray brown in dry, that has clusters of fuzzy white flowers and drinking goblet seed capsules.

lance-shaped, about 4 to 6 inches long by 1 or 2 inches wide and come to sharp points. The sickle-shaped leaves of eucalypts may resemble those of our native koa, but can always be told apart by the fact that these leaves will have a distinct mid-rib, while the koa phyllodes have parallel veins but no mid-rib. Leaf stems are often tinged with yellowish or pinkish hues. Flowers have 5 to 10 buds radiating from a point on a common stalk and with pointed, onion-dome caps that fall off as whitish, spreading filaments open up to form fuzzy, broadly conical flowers about 1¼ inch across. The seed capsules are shaped like miniature drinking goblets about ⅛ inches across and with 3 or 4 valves on top, that look like the Mercedes-Benz logo when viewed from above (when there are 3). This tree is native to coastal eastern Australia. It was first reported in Hawai'i from Lāna'i in 1916. Swamp mahogany has been planted here more than any other species of forestry tree, nearly 5 million having been set out before 1960.

FIDDLEWOOD
Citharexylum caudatum
Verbenaceae (lantana, vervain, beach vitex)

Fiddlewood was planted in what is now the Lyon Arboretum in 1931. It is native to the warmer parts of Latin America and the West Indies, and has taken well to life in Hawai'i, being readily spread by birds and is now a major weedy pest throughout the mesic and wet forests of the Ko'olau Range on O'ahu. This plant is a large shrub or small tree. Its paired leaves are opposite pointed oblongs about 2 to 6 inches long and 1 ½ wide and are well spaced along the twig. They are darker above than beneath and turn orange as they age. The tiny white flowers cling to the sides of a slender spike that is 4 to 8 inches long, and are followed by a cylindrical cluster of orange berries which turn black as they ripen, so that the cluster may present a mosaic of black and orange along its length.

KEN SUZUKI

A close relative of fiddlewood, *C. spinosum*, has been planted as an ornamental tree on all the major islands of Hawai'i, and appears to have a similar potential to become a serious invader of native forests. It is said to favor dry, sunny, rocky hills and to be well established on O'ahu. This tree is also called fiddlewood.

Look for: An upright shrub or small tree with opposite, well-spaced leaves, the older ones turning orange before falling, and slender spikes of small white flowers or orange berries, sometimes mixed with black.

FIRETREE
Morella (Myrica) faya
Myricaceae (wax myrtle, bayberry)

These weedy alien plants are shrubs to trees up to 30 feet tall or more, with closely spaced alternate oblong dark green leaves about 2 inches long that have edges that turn under slightly. The bark is gray to brown and smooth on younger trunks. Twigs are covered with rust-colored hairs. The leaves tend to be wider near the bluntly pointed tip and taper toward the base. Inconspicuous flowers give rise to clusters of ¼ inch diameter red to black berries which are eaten and dispersed by birds. The tree is native to the Canary Islands, Madeira, and the Azores and was probably introduced into Hawai'i by Portugese laborers as an ornamental and for the fruit which they used to make wine. It was used for forestry plantings in 1926 and 1927, but by 1940 it was considered to be one of Hawai'i's most noxious pests. It spreads through fields, pastures, and forests in mesic to wet areas on all the larger Islands crowding out desirable plants and forming dense, impenetrable thickets with such deep shade that few other plants can survive beneath them.

These trees form a nitrogen-fixing association with bacteria of the genus *Frankia* that permits them to grow in nutrient-impoverished volcanic soils. They enrich the soils beyond what native plants are adapted to, and allow other invasive weeds to colonize the area. As the name implies, they support wild fires but recover quickly after burning, allowing them to replace native vegetation destroyed by the fire. These trees are a particular pest in Volcanoes National Park on the Big Island where they have displaced native 'ōhi'a trees over thousands of acres of recent lava flows.

Look for: A tree of mesic to wet forest with gray to brown smooth bark and alternate dark green 2 inch leaves crowded at the twig ends, with edges that turn under slightly.

GUNPOWDER or CHARCOAL TREE
Trema orientalis
Ulmaceae (elm, hackberry)

These are weedy, fast growing trees that may reach 70 feet in height. The tree has smooth gray-brown bark and branches that tend to droop. It has well-

KEN SUZUKI

spaced alternate leaves that are elongated ovals tapering to a more or less pointed tip, and are 3 to 6 inches long. The leaves lie in a plane on opposite sides of the finely hairy twig. They are edged with fine teeth. The upper leaf surface is rough—slightly sandpapery to the touch, while the lower has prominent raised veins and is pale, relatively soft, and covered with a slight fuzz. Small flowers cluster at the base of the leaf and mature into small black berries less than ¼ inch in diameter. The tree is native to tropical Africa, Madagascar, south east Asia, Japan, Australia, and many islands of the Pacific, but not Hawai'i. Apparently, it is not known how this tree reached Hawai'i, but it arrived here sometime after 1870 and

Look for: A weedy, open tree of mesic forest with alternate leaves lying in a plane that have a sandpapery upper surface and a pale, soft under surface. The leaves are edged with fine teeth.

now occurs in mesic forest on all the main Islands. The light-weight wood is reported to make a good charcoal for use in fireworks and gunpowder.

HALA PEPE
Pleomele spp.
Agavaceae (ti, agave, New Zealand flax)

There are 6 species of this endemic native tree in Hawai'i, 2 of them found on O'ahu. This small tree looks like something from one of Dr. Seuss' books—with a somewhat spindly trunk topped by a tuft of slender, strap-like leaves hanging down from the tip of each branch. The plant resembles another native, the 'ie'ie, but the latter is a vine that clings to trees for support and has a prominent midrib in each leaf and small teeth along the edges of the leaves. Hala pepe in contrast is a free standing tree, up to 30 feet tall, with no obvious midrib in the leaves which have smooth edges. When in bloom, the tree has large clusters of yellowish flowers which are followed by red berries half an inch or more in diameter. The trees are found in the upper dry and mesic forest.

JOHN HOOVER

Look for: A tree with a tuft of long slender pointed strap-like leaves hanging down at the end of each branch.

A flowering branch of hala pepe was placed on the altar in the hula hālau to honor the goddess Kapo. Sometimes this plant has been considered to belong to the genus *Dracaena*.

HAME
Antidesma spp.
Euphorbiaceae (kukui, spurge, tapioca, castor bean)

There are 2 endemic species in this genus in Hawai'i. They are trees up to 30 or 40 feet tall, with alternate pointed broadly oval leaves about 4 inches long by 2 ½ inches wide. The leaf stems are short stout and strongly curved, forming a hook, that allows the leaves to lie in a plane along the twig. The small inconspicuous flowers are followed by clusters of fruit that are reddish to dark purple when ripe, round and flattened like small M & M's. These plants are scattered in mesic to wet forest on all the larger Islands.

KEN SUZUKI

Look for: Trees with alternate, pointed, broadly oval leaves with short, hook-like leaf stems, and, when in fruit, reddish to purple M & M-shaped fruit.

The hard red-brown wood of these trees was used by the Hawaiians for anvils on which to beat the olonā fibers that they used for their best nets and cordage. Poles were occasionally used in house construction, and the fruit yielded red to purple dye for staining kapa cloth.

'ILIAHI, SANDALWOOD
Santalum spp.
Santalaceae (I know of no familiar relatives)

Many species of this tree with its fragrant heartwood are found throughout India, southeast Asia, Australia, and the islands of Polynesia. One was even found in Chile, although it may now be extinct due to over-harvesting. Four endemic species are found in Hawai'i, 2 of which are fairly widespread. The trees are hemi-parasitic, meaning that their rootlets tap into those of neighboring plants and obtain part of their nourishment from them. The trees produce clusters of small

Look for: A shrub of dry coastal lowlands and adjacent ridges with silvery oval leaves, and small star-shaped flowers in season.

¼ inch diameter 4-petalled star-shaped yellow or red flowers and oval, olive-like fruit, with a cap at one end, about ¾ of an inch long. Do not break a twig or crush a leaf to try to detect the scent. Only the heartwood has the distinctive sandalwood fragrance. The 2 species of 'iliahi that you are most likely to see are:

S. ellipticum, or 'iliahialo'e, the coastal sandalwood. This is a shrub or small tree of low elevation dry areas. It has oval gray-green leaves 1 or 2 inches long. The largest one I know of on O'ahu is a bushy shrub about 10 feet high, but apparently the plant can form a small tree with a trunk a foot in diameter on the other Islands, according to some reports. The leaves of this plant can be quite variable, and I have often seen plants that appeared to be hybrids between this species and the next one. 'Iliahialo'e is found on all the main Islands.

KEN SUZUKI

S. freycinetianum, or mountain sandalwood, can grow to be a middle-sized tree up to 40 feet tall or more. Its leaves are spindle-shaped, about 3 inches

Look for: Shrubs or trees of the mesic to wet forest with star-shaped flowers, reddish new leaves, and droopy, folded, wavy older leaves.

long and an inch wide, usually with wavy edges that fold toward each other along the midrib. The leaves tend to hang down, giving the tree a somewhat droopy look. Emerging young leaves are usually erect and have a reddish tinge, which is a distinguishing feature once you are familiar with it. The tree is found in mesic to wet forest, on slopes and ridge tops, on all the major Islands except Hawai'i. It is not uncommon on mesic ridges on the leeward side of the Ko'olau Range on O'ahu, and may extend down to overlap the upper range of the preceding species.

The 3rd Hawaiian species is found only on Haleakalā on Maui while the 4th is scattered in dry woodland on the Big Island. The Hawaiians had limited uses for 'iliahi, using it to perfume kapa by sprinkling the powdered heartwood between sheets of the cloth or mixing it with the coconut oil used to waterproof the material. The bark may also have been used in preparations intended to sooth aching joints.

'Iliahi played a major role in the early westernization of the Hawaiian economy and the destruction of its traditional culture. Sandalwood had been known to the Chinese for centuries and they used the fragrant wood for furniture, chests, cosmetics, incense, medicines, insect repellant, and perfume. I have heard that the very wealthy would even flaunt their prosperity by being cremated on a pyre of sandalwood. Most Chinese sandalwood was imported from India, but supplies were limited and prices very high. In New York, this wood commanded a higher price than any other. An American fur trader, Captain John Kendrick, from Boston, apparently discovered the presence of sandalwood in Hawai'i in 1791 and began the trade with China. As long as Kamehameha I reigned the trade was somewhat controlled, but after his death in 1819, by which time the ali'i had discovered the wonderful (and highly over-priced) luxury goods that they could buy with the money from the trade, a virtual frenzy of exploitation of this tree began. In some areas, commoners, including women and children, were driven into the mountains, with little food or protection from the unfamiliar cold and rain, and forbidden to come home until they had fulfilled their quota of logs. They were forced to dig out even the roots of the trees to satisfy the demands of their Lords. The neglect of their crops led to at least 2 famines during this period, and the stress must have been a factor in weakening their resistence to the numerous alien diseases that increasing trade with both East and West introduced to the Islands and that had such devastating effects on the population. The commoners are said to have hated the trade so passionately that they would pull up any sandalwood seedlings they saw, in the hope of sparing their children from similar labor. Forests were often burned to help locate the fragrant trees, making passage through the area easier, but destroying the native ecosystems and stripping the watershed of all protection. By 1830, the sandalwood trees had been seriously depleted, and so few were left by 1840 that the trade came to an end.

KOA
Acacia koa
Fabaceae (wiliwili, royal poinciana, peas and beans)

The koa is the largest and second most common of our native trees. ('Ōhi'a lehua is the most common). In favorable situations, koa may grow to be a tall, straight tree of great height and girth. The largest double canoe ever reported from Hawai'i was made from koa trees harvested in Kīpahulu Valley on Maui, and the canoe bodies were 120 feet long and 9 feet deep, each carved out of a single huge koa log. Such trees were always very rare, however, and practically all the large koa that you are likely to see are gnarled old trees that have massive, low branches and a wide crown, but no tall straight trunks. Koa first appears in the upper dry forest and extends upward into the wet forest. Young trees have smooth, light gray bark, but this becomes dark, rough, and furrowed as the tree ages. The bark on a young koa may look blotchy due to the growth of light colored lichens, or may even have a reddish tint where a green alga, *Trentepohlia aurea,* grows on it. The koa can easily be recognized by its sickle-shaped leaves. At first glance, these may look like eucalyptus leaves, but on closer examination you will see that while eucalyptus has normal leaves with a mid rib running down the center of the leaf and side veins branching off from it, the koa leaf has no mid rib or side veins, but a lily-like set of parallel veins running the length of the leaf. These are not actually leaves at all, but something that botanists call "phyllodes" which are the stems of a leaf that have become broad and flat and taken over the functions of a leaf. For simplicity, I'll just call them leaves, however. As the scientific name implies, the koa is an acacia. Young seedlings, and the water shoots that emerge from an older tree at the site of an injury, show the true leaves. These are typical acacia-style doubly compound leaves with pairs of stalklets, emerging from the leaf stalk at intervals, that bear pairs of oval leaflets along their lengths. Frequently the leaf stalk is beginning to broaden, and you can observe an intermediate between the true leaf and the sickle-shaped replacement form seen on the mature tree. It is an interesting experience to see a little tree bearing two such dramatically different kinds of leaves, as many young saplings will have both sickle leaves and acacia-like true leaves, as well as some which haven't quite decided which form to take. The only tree that you are likely to confuse with koa is its close relative, the Formosan koa. This tree does not get as large as our native koa. Formosan koa never produces true leaves, and its phyllodes are smaller and less curved and have blunter tips than in true koa. They tend to stick straight out from the twig so that their tips describe (roughly) a cylinder, while the sickle leaves of koa tend to hang down, so that the blade of the leaf lies in a vertical plane which minimizes its exposure to sunlight when the sun is at its highest in the sky. This may be an adaptation to dry conditions and help to reduce water loss. In contrast, the true leaves tend to lie in the horizontal plane, and catch the sun's rays. Koa flowers are small round puff balls, like those of its Formosan cousin, but a paler yellow in color. They are followed by bean pods. Koa not only reproduces by means of its seeds, but also sends up shoots from its roots. This helps to make it one of the few native plants that rebounds rapidly after a fire, since many suckers sprout up from the roots even though the mature tree has been killed.

The Hawaiians used koa wood for many purposes. It was the major source of canoe logs, and an elaborate ritual was required before a giant koa could be harvested for a large double canoe. The harsh cry of the ʻalalā, the Hawaiian crow would indicate that the day was not auspicious for cutting the tree. Once the tree was felled, it was left in place until Lea, female deity of canoe builders, in the form of the ʻelepaio, a small, friendly native bird, had inspected it and determined that the log was sound. If the ʻelepaio stopped and pecked diligently at the log, indicating an abundance of insects and perhaps decay, the log was rejected. Koa was also used for canoe paddles and the short form of surf board. It was made into bowls for storage purposes, but not used where food would come into contact with the wood, because a resin in the wood imparts a bitter taste to food. A red dye was obtained from koa bark. When the Portuguese arrived in the Islands, the first ukeleles were made of koa wood.

Koa may be one of the most ancient arrivals in Hawai'i, since nearly 50 species of insects have become adapted to living on this tree, more than on practically any other native plant. Another recent finding of interest is that koa is associated with two different forms of the nitrogen-fixing bacterium, *Bradyrhizobium*. This is the organism that often forms nodules on the roots of leguminous plants—members of the pea and bean family to which koa belongs. One variety of *Bradyrhizobium* forms such nodules on the roots of koa while second one forms nodules on adventitious roots in pockets of soil and leaf debris in the crotches of koa branches or in hollows along the branches where such material collects. This habitat may be exceptionally favorable for the bacterium since it is rich in organic nutrients and low in the aluminum ions that are common in tropical soils, but are rather toxic to many organisms. Such canopy-associated nodulation is known in only one other species of tree.

JOHN HOOVER

Look for: Large, gray-barked trees with sickle-shaped leaves, lacking a midrib, that hang downward. Young plants will have true, acacia-type leaves instead of, or in addition to, these.

KŌPIKO
Psychotria spp.
Rubiaceae (coffee, quinine-yielding Cinchona, gardenia, alahe'e)

The kōpiko are attractive, common trees or shrubs with glossy pointed oval leaves. They are found in the upper mesic zone and extend into the wet forest. Eleven endemic species are found in Hawai'i, but they are quite similar to each other. They can easily be recognized by two features. The first is the presence of piko (navel, in Hawaiian) or small pits that lie in the corner between the mid-rib and side veins on the underside of the leaf. These are often quite prominent, but may be small or even absent in some species. Scientifically, they are called "domatia" (as in domicile, or home) and house bacteria of the genus *Azotobacter* which are able to fix nitrogen. All organisms need nitrogen-containing compounds, but only bacteria (and humans, in fertilizer factories) are able to perform the difficult, energetically expensive

123

job of capturing it from the air. Presumably, the plant supplies the bacteria with sugars produced by photosynthesis, and benefits from the nitrogen-containing compounds in return.

A second, even more reliable feature of the kōpiko, is the flower stalk. This is usually erect, although it may droop in some species, and bears small, white, star-shaped flowers with 4 or 5 petals. The flower stalk branches in a very characteristic pattern. Four branches emerge from the same point on the straight stalk at roughly right angles to it and to each other forming a cross-like structure resembling a child's jacks. Additional stalklets occur at the top of the stalk and ends of the side branches, and the flowers are found on these. Once seen, this structure can never be mistaken for any other, and clearly identifies the kōpiko. The flowers are followed by oval orange fruits about ½ inch long. Kost Pankiwskyj tells me that these are sometimes almost palatable to eat, though like all fruits, they vary in quality from one to another. Never eat any unfamiliar fruit unless you are absolutely sure that it is safe, of course.

JOHN HOOVER

JOHN HOOVER

Look for: A tree with glossy leaves that have pits along the midrib on the underside of the leaf, and a jacks-like structure of the flower stalk.

Kōpiko are widely distributed through the tropical world—there are about 1,500 species, altogether! I once joined an Earthwatch project in Cameroon, working with botanists from Kew Gardens to collect and preserve plants in an area that we hoped could be protected as a nature reserve. There was a *Psychotria* there (they wouldn't have known what I was talking about if I'd called it a kōpiko) that had enormously long flower stalks—several yards in length, that hung down to the ground. They were red and fairly thick, and the tree looked as if someone had draped it with electrical cables. Internet connections, no doubt. I surmise that it was pollinated by some ground-dwelling beetle or similar insect, and had evolved to bring the flowers close to the pollinator. I have found no mention of any use the Hawaiians had for this attractive and common native tree.

KUKUI, CANDLENUT
Aleurites moluccana
Euphorbiaceae (spurge, manioc, poinsettia, castor bean)

The kukui was introduced to Hawai'i by the earliest Polynesian settlers, sometime between 2000 and 1600 years ago. It is a common tree in moist gulches at relatively low elevations throughout the islands. This tree is a fairly large, spreading tree with grayish-white bark and pale green leaves that are often 10 inches or more in length and have pointed lobes, much the same shape as a maple leaf.

KEN SUZUKI

Look for: A tree with large pale maple-like leaves with 3 to 5 pointed lobes and numerous gray or dark brown nuts on the ground.

The leaves, especially when young, are frequently hairy, particularly on the bottom side. Kukui trees bear clusters of small white flowers followed by large nuts which appear first as a round fruit about 1 ½ inches in diameter. The husk rapidly decays after falling to the ground, leaving a hard-shelled nut about 1 inch in diameter. The nut is flattened on one side, with a slight groove down the center, while the other is domed with irregular shallow furrows. They can be a hazard on the trail as they may roll underfoot like ball bearings. It is easy to recognize kukui trees in gullies in the distance, since their pale leaves stand out among the darker foliage of other trees.

The Hawaiians had many uses for this tree. Roasted nuts were strung on the midrib of a coconut leaflet and used as candles—hence the English name, "candlenut tree". Or the oil could be expressed and burned in stone lamps with a kapa wick. The oil was used to give a gloss to bowls, canoes, and surfboards. The raw nut is a powerful laxative—the saying is that if you eat one, you walk to the nearest lua. If you eat two, you run. And if you eat three, it is too late! However the roasted nut was mashed and mixed with sea salt and the dried ink sack of a squid (chili peppers are often substituted today) and eaten as a condiment with food. Fishermen also chewed the nuts and spat them on the water so the oil would create a smooth window and allow them to see the hiding places of octopuses and fish. The leaves, flowers, and nuts were all used in different kinds of lei. Juice from the inner bark of the root served as the binder in paint made from finely powdered charcoal that was used to stain canoes, and the soot from burned nuts formed a black pigment, for tattooing, among other uses. A reddish dye obtained from the bark of the tree tanned and preserved fish lines and nets and made them less visible to the fish. The immature fruit yielded another dye that gave kapa a beige color. Kukui wood was used for components of canoes.

The tree had almost as many medicinal uses as the noni. Different parts of the plant were used alone, or in combination with other herbal concoctions to treat chills, constipation, asthma, thrush, stuffy noses and sinus problems, sores, rheumatism, deep bruises, sore throats, rashes, and chapped lips. A word of caution: The statement in books of this kind that a preparation was used to treat a malady does not necessarily imply that the treatment was effective, or if so, that it was due to the properties of the plant! We do not know exactly how many of these treatments were prepared and most of them involved mixtures of several different plant extracts and other materials, such as sea salt. Invariably, in important cases, they were accompanied by prayers, chants, purges, incantations, and ceremony. For those who had faith in the power of the kahuna, the psychological impact must have been substantial, and while some of the herbal remedies might well have had significant physiological activity, others may have depended mainly on the comforting reassurance that some powerful forces were being enlisted to restore health.

The kukui was regarded as one of the plant forms, or kino lau, of the pig demigod, Kamapuaʻa, who in turn was sometimes regarded as an incarnation of the major god, Lono. If you take a kukui leaf and fold it in half, with the crease running from tip to stem, you will see the profile of a pig—Kamapuaʻa! I have heard a legend that in the very beginning, before there was any sun, only the moon shone on the land. Lono, god of the forest (among other things) was unhappy because there were no trees. So he took a bit of clay and rolled it into small balls to make seeds, and planted them to start a forest. Unfortunately, with no sunlight, most of the trees could not survive, but the kukui, with its pale leaves, could capture the moonlight, and so was able to grow.

MACADAMIA
Macadamia integrifolia
Proteaceae (silk oak, proteas, banksias)

There are 10 species in this genus, 6 being endemic to Australia, 1 to Sulawesi, and 3 to New Caledonia. They are valuable timber trees and 2 of the Australian species bear edible nuts. The species above has been widely cultivated in Hawaiʻi for many years. The tree grows to be 60

KEN SUZUKI

feet tall. It has leaves in whorls of 3. The leaves are oblong, bluntly pointed, and about 5 inches long by 1 inch wide, with widely spaced spines along the edges. Small white flowers are borne along the sides of a dangling spike. The nut is about ½ inch in diameter and enclosed in a smooth, hard, spherical shell that is surrounded by a fibrous husk. It is thought that the first trees were introduced to Hawai‘i before 1837 by Don Francisco de Paula Marin, who brought so many useful plants to the Islands, along with some that have proved to be pernicious weeds. The trees were planted in orchards and for reforestation and are not infrequently seen along our mesic forest trails, on O‘ahu, at least.

Look for: Trees with leaves in whorls of 3 that are about 5 inches long, bluntly pointed, and usually with sharp spines along the margins. Nuts or husks on the ground are often noted as the first evidence of their presence.

MACARANGA
Macaranga mappa and *M. tanarius*
Euphorbiaceae (poinsetta, kukui, castor bean, tapioca)

These two trees are unmistakable because of their unusual leaves. These are shaped like broad upside-down tear drops, and like the taro, but unlike most other plants, the stem is attached to the leaf, not at one edge, but to the upper part of the back of the leaf. Both have large leaves, but those of the first, *M. mappa*, are particularly large, often being well over a foot long. The tree itself is a leggy, sparsely branched plant up to 30 or 35 feet tall. It is native to the Philippines and is now found at low elevations in mesic to wet areas in Hawai‘i. It was first collected here in about 1927.

M. tamarius has somewhat smaller leaves and a more typical tree form. It came from the islands of the southwest Pacific and is naturalized in disturbed mesic valleys in Hawai‘i. It was first noticed here in the wild in 1930. The tree also grows to 30 or 35 feet.

KEN SUZUKI

In their native lands, some species of macaranga play host to ants, which they encourage to nest in the tree by providing nest sites in pouch-shaped structures at the base of the leaves, or in hollow branches. The tree also furnishes nutritious bodies containing starch, fat, and some protein for the ants. In return, the ants attack insects and other herbivores that might damage the tree.

Look for: Trees with large, broad, tear-drop-shaped leaves in which the stem attaches to the upper back of the leaf rather than to an edge.

MANGO
Mangifera indica
Anacardiaceae (Christmas berry, sumac, cashew, pistachio, poison ivy)

Mango is a large spreading tree that is common in gardens in Hawai'i and should be familiar to most people. The crown is rounded and dense with oblong sharply pointed dark green alternate leaves that are about 6 to 12 inches long and 1 ½ to 3 inches wide. The flush of new leaves appears as a drooping cluster of pinkish-orange foliage. In late winter to early spring, large clusters of small yellow green to pink flowers are produced. These are followed by

KEN SUZUKI

Look for: Large trees with a rounded crown that cast dense shade and have long, slender, oblong, sharply pointed leaves, found near former house sites.

large oval fruit. A good mango is one of the most delicious of all fruits. The tree has been cultivated in India for about 4000 years and many horticultural varieties are available. It was probably introduced into Hawai'i as part of the shipment of plants on HMS Blonde that arrived here in 1825. Several of these trees were apparently planted by Don Francisco de Paula Marin. Along our trails, the variety of mango that is usually seen is that known as "common mango" and is ordinarily encountered in mesic valleys near abandoned house sites. Many of these trees are quite large—4 to 6 feet in diameter or more.

Mango is in the same family as poison ivy and poison oak, and people who are sensitive to these plants may develop a similar rash on contact with it. We had a lovely big Haden mango in our front yard for many years. Early one spring I visited a friend in New Jersey who had just bought a new house and helped him clear some vines off his trees. The leaves were not out yet, and we did not realize that some of the vines were those of poison ivy, a fact that became apparent when I developed a severe itchy rash during the flight home. After that, the rash tended to recur every time I came in contact with the mango tree, and eventually I had to have the tree removed.

MAUA
Xylosma spp.
Flacourtiaceae (chaulmoogra, Ceylon gooseberry)

There are 2 endemic species in this genus in Hawai'i. They are small trees, to about 30 feet tall, with alternate, pointed, broadly oval leaves about 4 inches long by 2 ½ inches wide, usually with blunt teeth along the edges. The young leaves and leaf stalks are often red in color. Clusters of small inconspicuous flowers are followed by oval reddish-purple fruit about ½ inch long. The trees occur occasionally in mesic forests on all the larger Islands.

The leaves of maua look very much like those of the unrelated and commoner hame, but the young leaves of the latter are not usually red and its fruit is shaped like a somewhat flattened disk.

Look for: A tree with alternate, pointed, broadly oval leaves with blunt teeth and a fruit that is oval, shaped like a football with rounded ends.

MOHO, WHITE MOHO
Heliocarpus popayanensis
Tiliaceae (blue marble tree, kalia, jute)

Look for: Trees with broad, alternate, often duck's foot-shaped leaves, and in fruiting season, tufts of fluffy lavender capsules in the tree tops.

This is a fairly tall tree, up to 100 feet in height with alternate leaves that are heart-shaped or with 3 points like a fat duck's foot. Leaves are 8 or 10 inches long and nearly as broad, and are hairy underneath. Clusters of pale yellow 4 part flowers are followed by small flattened capsules surrounded by a sphere of feathery hairs. From a distance these look like tufts of lavender blossoms on the top of the trees. These seeds are readily spread by the wind, permitting the tree to invade our native forests, where it threatens to become a serious pest. Moho is native to the Americas, from southern Mexico to Paraguay and was introduced to Hawai'i at the behest of the Hawaiian Sugar Planter's Association in the 1920s for reforestation. About 25,000 trees were planted on Kaua'i, O'ahu, and especially Hawai'i and have since spread.

MOUNTAIN APPLE, 'ŌHI'A 'AI
Syzygium malaccense
Myrtaceae (allspice, Java plum, guava, eucalyptus, 'ōhi'a)

This Polynesian introduction is a tree of low elevation mesic to wet valleys. The large opposite glossy green pointed oval leaves are 6 to 14 inches long by 2 to 8 inches wide. The strikingly beautiful flowers are tufts of stamens, like those of the related 'ōhi'a, crimson in color with just a hint of lavender. They appear along the sides of twigs and branches. Blooming is somewhat unpredictable and may occur in Spring and Fall. It is followed by 2 to 4

Look for: Groves of trees up to 50 feet tall in mesic valley bottoms with large glossy leaves and lovely pompom red flowers or pear-shaped fruit in season.

inch-long fruits which are pear-shaped with a large dimple in the bottom. When ripe these will be red to maroon in color and are crisp, sweet and juicy but without a distinctive flavor. A single large round seed occupies a cavity in the center of each. A few fruit can be eaten with relish on the trail, but eating too many can lead to stomach ache, and the Hawaiians often processed them by splitting the fruit and partially drying the halves on skewers before consuming them. The native range of this tree is probably south east Asia and the adjacent islands.

The Hawaiians had a number of uses for this tree. The fruit was eaten, and the wood was used in house construction and for temple enclosures, and idols were carved from it. Medicinally, leaf buds were used in a mixture with other plants to treat chills and the bark was pounded to obtain an extract that was combined with other herbals in a variety of concoctions to treat chest pain, asthma, tuberculosis, abdominal ailments, sore throat, bronchitis, and, as a gargle, for bad breath, colds, and sinus problems. A black dye for kapa was obtained from the inner bark of trunk or root, and a red dye from the skin of the fruit.

OCTOPUS TREE, RUBBER TREE
Schefflera actinophylla
Araliaceae (panax, English ivy, ginseng)

These trees have light tan bark and are up to 45 feet tall with thick shoots and twigs. The large leaves are palmately compound with the main leaf stem being 7 to 18 inches long and with 5 to 18 large oval leaflets, each 4 to 18 inches long, attached to a common point at its end on 1 to 3 inch long stemlets. Small reddish flowers cluster along stiff, spreading stalks about 2 feet long during the fruiting season from April to October. Five or 6 such fruiting stalks radiate from a point near the end of a branch and resemble the outstretched arms of an octopus, giving the tree its common name. This plant is native to Australia and New Guinea. It has been grown in Hawai'i as a garden and pot plant since about 1900, but is now naturalized and threatening to become a serious invasive pest in mesic and wet forests on all the major Islands. The seeds are readily spread by birds. It does not yield rubber.

JOHN HOOVER

Look for: A tree with light tan bark and large palmately compound leaves. During much of the year, octopus-like red colored arms may be seen on its top.

'ŌHI'A, 'ŌHI'A LEHUA
Metrosideros polymorpha, Metrosideros spp.
Myrtaceae (guavas, eucalyptus, cloves, allspice, mountain apple)

'Ōhi'a is our commonest native tree, and one of the largest. There are 5 endemic species in the genus, and as the scientific name implies (poly = many, morpha = forms) the commonest species is quite variable in its appearance. The early Hawaiians recognized the different species, and had distinct names for most of them. The plant is usually

KEN SUZUKI

a tree, but may occur as a shrub in dry areas or even as a dwarf a few inches high, yet completely mature and flowering, in a bog. It is found in the mid to upper dry forest and extends into wet cloud forests and grows from sea level on windward coasts to timberline on the borders of the alpine zone at an altitude of about 8500 feet on our high mountains. The bark of the tree is light gray, rough, fissured and scaly, though in moister areas it may be obscured by growths of lichens, liverworts, and mosses. The leaves are opposite, simple, generally small, from ½ to about 2 inches in length and very variable in shape and texture. Alternate pairs of leaves are arranged quite regularly at right angles to each other. Many 'ōhi'a leaves have galls on them. These are small round colored lumps on the leaf surface. They are caused by one of perhaps 10 different species of a tiny native insect called a "psyllid". This insect lays an egg in the leaf tissue, and a larva hatches and produces some chemical that the plant recognizes as a hormone and responds by producing the abnormal growth resulting in the gall. The larva feeds on sap from the swollen leaf tissue until it is mature and ready to emerge. Then it exits the gall through a tiny hole, molts into the adult form, and starts the cycle over again. Similar galls on alani and pāpala kēpau trees are caused by other native psyllid species.

JOHN HOOVER

The most significant feature of 'ōhi'a in terms of recognition is the blossom. This is a red pompom with inconspicuous petals but long red stamens. (Lehua—the name of the

Look for: An often spindly, rather gangly tree with rough, fissured gray bark, small leaves, and red pompom-like flowers.

blossom–means "hair" referring to this filamentous flower.) An occasional tree will have yellow or salmon colored flowers, but most are red. While not every tree you meet will be in bloom, there will almost always be some blossoms somewhere along the trail, and once you have recognized the tree, it should be easy to identify others that are not in flower. The 'ōhi'a has the most conspicuous and colorful blossoms of any native tree, and an 'ōhi'a in full bloom is indeed a glorious sight!

The 'ōhi'a is a pioneer species, and on the Big Island it is one of the first plants to grow on a fresh lava flow, often sprouting up out of cracks in a smooth pāhoehoe surface, that would not seem to offer any possibility for plant growth. Often an 'ōhi'a tree will send down aerial roots, either as single filaments or massive clumps of hundreds of rootlets—a "beard." The function of these is not well understood. In cloudy areas, they may help to capture droplets of fog to water the tree. When 'ōhi'a trees were buried by 4 or 5 feet of ash in the Kīlauea Iki eruption, many were able to survive by re-rooting in the fallen ash, at a higher level, by means of these aerial roots. Plants that could not do this died when their roots were smothered by being buried so deeply. When the aerial roots reach the ground they take root and the tree acquires a prop, another trunk, to brace it. Some 'ōhi'a trees appear to be standing on stilts, although this probably happens more often when seedlings germinate on the trunks of a tree fern or decaying nurse log. The seedling sends its roots around the log or fern to the ground, and when the support has decayed away, the tree is left standing on its roots, with its trunk beginning well above the ground surface. Since so many 'ōhi'a get their start on a fern trunk, the Hawaiians regarded the fern as the "parent" of the 'ōhi'a. 'Ōhi'a roots also support an unusual ecosystem. Only a few years ago, Frank Howarth, an entomologist with the Bishop Museum, discovered that lava tubes have their own unique suite of cave-dwelling animals. These are invertebrates that live in the perpetual dark, feeding on debris that washes into the lava tube and on each other, and especially on the roots of 'ōhi'a trees that can penetrate for more than 30 feet through soil and rock to emerge from the roof of the cave.

'Ōhi'a are also important as a food source for native birds. Honey creepers and other forest birds forage around the blossoms for nectar and the small insects to be found there. Sometimes there are many 'ōhi'a in bloom, sometimes few. It seems likely that many birds may have migrated seasonally, or even daily, between these forests and other food sources in the lowlands. One source speaks of how the 'ō'ō and 'i'iwi were trapped near the mouth of Pu'uloa (Pearl Harbor) on O'ahu when they came down to feed on ripe noni fruit. 'ō'ō, now extinct, furnished small tufts of yellow feathers for the yellow capes and helmets of the ruling ali'i, while 'i'iwi supplied red feathers for the same artifacts. Such migrations may have proved fatal to the birds once whaling ships introduced mosquitoes and foreign birds brought in the bird malaria and bird pox, to which the natives had no resistance. The mosquitoes, which transmit these diseases, do not occur high in the mountains, but if the birds periodically came to lower elevations, they could have been infected there. These diseases are thought to be a major factor in the extinction of so many of our forest birds since the arrival of Captain Cook in the Islands. Many native plants—a variety of lobelias, mints, and the 'ohe naupaka, for example—have long, curved blossoms that were probably pollinated when the long, curved bills of some of these birds were used to probe them for nectar and insects. When the bird becomes extinct, what happens to the plants that it pollinated? Auwe! Auwe!

Because it was so common, 'ōhi'a was used for most purposes which required hard, strong wood, although it cracks too easily to be suitable for bowls. Red was the sacred color and since the heart wood of 'ōhi'a has a reddish tinge, this wood was preferred for the carving of the large images that adorned major heiau (temples). Perhaps for this reason, the tree was believed to be a kino lau of both Kū and Kāne, two of the four most important deities in old Hawai'i. (The Hawaiians believed that many natural objects—rocks, plants, and animals—were *kino lau*, or body forms of the gods. Each of the major gods had a number of kino lau in which he was simultaneously present.) Smaller, portable images, or akua kā'ai, were also usually made of 'ōhi'a.

'Ōhi'a and lama woods were used in construction of the sacred buildings within a heiau. Since Laka, patron goddess of hula, was also the forest goddess, a flowering branch of 'ōhi'a lehua was always placed on her altar in the hālau hula. 'Ōhi'a wood was used for the gunwales of canoes, house rafters, scraping boards for preparing wauke bark for making kapa cloth, and many other purposes. The leaves, flowers, and leaf buds of 'ōhi'a were used in lei. The tuft of new leaves at the end of an 'ōhi'a twig is often a bright red or furry white in color and these, which are called "liko", are still commonly gathered and used in haku lei.

Hawaiians have a belief that picking lehua blossoms will bring rain. Therefore, they should be picked while coming out of the mountains, not while going in. This is explained by the idea that the lehua was sacred to Pele, and if she was offended, she might send mist and rain to envelop the party.

'ŌHI'A HĀ
Syzygium sandwicensis
Myrtaceae ('ōhi'a, mountain apple, eucalyptus, guava)

When I first started learning to recognize the native trees, I had great difficultly in distinguishing this endemic tree from 'ōhi'a, to which it is related. The leaves are similar in size and shape to those of certain varieties of 'ōhi'a, being oval with bluntly pointed tips and about 1 to 4 inches long by ¾ to 1 ½ wide. The young leaves have a slightly ruddy tinge, and later leaves are somewhat yellowish, only darkening to a mature green as they age. Thus leaves at the twig tips are lighter than those further

Look for: A tree of mesic to wet ridges and slopes with 'ōhi'a-like leaves, patches of guava-like bark, and small red berries. Note the leaf emergence pattern.

down. Leaf emergence in the two species is quite different, as well, with a flush of many paired leaves in 2 rows being produced at once in the 'ōhi'a while 'ōhi'a hā sends out just one pair of slender leaflets at a time. My friend, and expert amateur botanist, Ken Suzuki, says that the tree can be recognized if you crush a leaf and note the distinctive fragrance. My nose is poor and I cannot detect the odor, but many of you may be able to. 'Ōhi'a hā can get to be quite large trees with thick trunks and spreading crowns, although usually they are smaller and are often seen as shrubs. The bark is smooth and grayish brown and frequently has patches of smooth red-brown bark like guava, very different from the gray brown flaky bark of 'ōhi'a. Flowers are inconspicuous small white blossoms with 4 tiny petals, and are followed by clusters of red berries about ⅓ inch in diameter. These are edible, although there is only a thin layer of flesh around the seed in the center. The tree can be considered a native version of its close relative, the mountain apple, or ʻōhiʻa ʻai, which the Polynesians introduced to Hawaiʻi, and which has a much larger, more satisfying fruit. ʻŌhiʻa hā is found on ridges and slopes in mesic to wet forest on all the larger islands except Hawaiʻi.

ʻŌhiʻa hā is much less common than ʻōhiʻa, but by no means rare. The Hawaiians used the wood for fuel and house construction, and sometimes made canoe paddles from it. They obtained a black dye for staining kapa cloth from the bark.

PĀPALA
Charpentiera spp.
Amaranthaceae (pigweed, cockscomb, amaranth, kuluʻi)

There are 5 endemic species of small trees in this genus. They range from 15 to 40 feet in height and have large smooth alternate oval leaves, 6 to 12 inches long. The leaves look similar to those of pāpala kēpau, a tree of an entirely different family, which occurs in the same habitat, but if you feel the edge of the leaf it is usually possible to detect a raised margin on the pāpala leaf which is absent in pāpala kēpau. Young leaves are often slightly to densely hairy. These trees are common in mesic to wet gulches. Their relationship to other amaranths can be seen in the long thread-like branched fruiting structures that bear numerous tiny flowers and seeds.

The wood of pāpala is light and spongy, like balsa wood. On dark moonless windy nights, aliʻi would gather in their canoes in a few favored locations, such as Kēʻē Beach on Kauaʻi where young men would build a fire on top of Makana Promontory, which drops in a sheer cliff 1280 feet to the beach below. They would rub dry branches of the pāpala with kukui nut oil, ignite them, and throw them toward the sea, where the light wood would rise and fall with the wind as it floated downward, all the while burning and sending out jets of sparks from the soft, highly inflammable pith in the center of each branch. This was the Hawaiian version of Fourth of July fireworks! Sometimes a young buck in a canoe would catch a falling branch to demonstrate his dexterity, and brand himself on the chest to show his ardor for some damsel.

Thomas H. Rau

Look for: A tree of mesic to wet gulches with large oval leaves with a raised margin and (in season) long thread-like branched fruiting structures.

135

PĀPALA KĒPAU
Pisonia spp.
Nyctaginaceae (four-o'clock, alena, bougainvillea)

There are 5 species in this genus that are native to Hawai'i, 3 indigenous and 2 endemic. All are large shrubs to trees, with soft, brittle wood. The large leaves are opposite or in whorls. They are 4 to 12 inches long by 3 to 6 inches wide, leathery ovals with blunt tips. Small galls produced by a native psyllid insect are often found on the leaves (see 'ōhi'a). The flowers occur in large open clusters and are small, funnel-shaped, with 5

BRADLEY F. WATERS

Look for: A valley bottom tree of dry and mesic forest with large blunt oval leaves, with no raised rim, and very sticky cigar-shaped fruit.

lobes, and white, often with a pinkish tinge. The fruit is cigar-shaped, about 1 ½ inch long, and very sticky when ripe. The trees resemble pāpala, which is in an entirely different family, but the flowers and fruit are quite distinct. If no flowers or fruit are present, it is usually possible to tell the 2 trees apart by feeling for the slightly raised rim around the edge of the pāpala leaf, which is characteristic of this plant and not found in pāpala kēpau. *Pisonia* is a pantropical genus, primarily from tropical America, with a few species in the Pacific islands. It is generally found in lowland dry to mesic valleys in Hawai'i, on all the major Islands.

Insects are often trapped on the sticky ripe fruit and these attract birds which may also become caught by the glue. I recall a hike in the Wai'anae Range on O'ahu when we came across a young Brazilian cardinal that was unable to fly because a pāpala kēpau fruit had glued one wing to its body. We were able to capture the bird and free it. Such observations undoubtedly gave the early Hawaiians the idea of using this sticky substance to catch native forest birds that furnished the feathers for the spectacular capes, helmets, feathered gods, and lei that they produced. Hawaiian ali'i, like nobility everywhere, sought ways to demonstrate their wealth and status through the ostentatious display of rare and beautiful objects. Lacking access to precious metals and sparkling gem stones, they turned to objects created from the beautiful feathers of native birds. In other islands of Polynesia, red feathers were rare and thus were most valued by the aristocracy. However two of the most common forest birds in Hawai'i have lovely red feathers, while golden yellow feathers were found only in small tufts on two otherwise drably colored, less common, birds. Auwē!, both of these are now extinct. Thus in Hawai'i, while red continued to be the color favored for religious objects and sacred to the gods, capes of yellow were reserved for the royal paramount chiefs, and lesser aliʻi were permitted to include red, but only a smattering of yellow, if any, in their garb. The sticky glue from pāpala kēpau fruit, or the sap of ʻulu (breadfruit), was smeared on twigs or poles on which the birds were likely to light and to which they would then become stuck. Snares or mist nets of olonā fiber were probably also used to capture the birds which bore the desired feathers, and such feathers made up a major part of the annual tribute due the principal aliʻi of the island from districts, ahupuaʻa, containing substantial populations of these birds.

PAPERBARK
Melaleuca quinquenervia
Myrtaceae (eucalyptus, rose apple, ʻōhiʻa)

This is a tree growing up to 60 or 75 feet tall with thick spongy white or pale brown peeling bark. It is closely related to eucalyptus. The leaves are alternate, spindle-shaped and about 2 to 4 inches long. They are stiff and have no obvious mid rib, but with 5 or more parallel veins running from base to tip somewhat like the veins in the phyllodes of the koa. This tree cannot be confused with either of the koa species here, however, because of its characteristic thick spongy bark. Bottle brush-like white cylindrical flower clusters grace the ends of twigs in season. These are made up of numerous fine filaments (stamens). They are followed by many tightly packed small round woody cup-shaped seed capsules that surround

Look for: A tree with thick spongy peeling white bark and small spindle-shaped leaves with no mid rib.

and cling to the twig. The tree is native to eastern Australia, New Guinea and New Caledonia. Over 1.7 million of these trees have been planted in Hawaiʻi since seed was introduced from Miami, Florida in 1920. It is naturalized in disturbed mesic forest here now, but as yet has not become a serious pest, as it has in the Everglades and surrounding areas in Florida. The bark provides excellent insulation against fire to the living tissue beneath, permitting the tree to survive most brush fires. Heat causes the capsules to open and the wind that often accompanies a blaze helps spread the seeds. The relative infrequency of fires in our forests may explain why the tree has not spread here as successfully as it has in Florida.

PARASERIANTHES, "ALBIZIA"
Paraserianthes falcataria
Fabaceae (monkeypod, sweet pea, mesquite)

KEN SUZUKI

Look for: A large, white-barked tree with doubly compound leaves, in low elevation valleys and along old ditch trails.

I have always known these trees by the name "albizia" but the latest edition of the *Manual* says they are now in the genus *Paraserianthes*. I suppose that we will all eventually become accustomed to the new name—which is actually rather melodious. I just hope the botanists don't change it again, then. It is not unknown to have a name changed, only to have a later student of the group decide that the change was not really justified and change it back again. Or worse, to have the plant reassigned to yet a third, different, genus. It is enough to make one sympathize with those folks who want to abolish all scientific names and just assign each organism a number, a kind of bar code, which it will possess for ever and ever and never change, and which can be moved from one scientific pigeon hole to another as the facts seem to demand, without the bother of changing a name.

These trees are some of the largest to be seen in Hawai'i, growing to heights of 100 feet or so, with smooth white-barked trunks that are often 4 feet or more in diameter, and often have prominent spreading above-ground roots. The leaves are doubly compound with 8 to 15 pairs of stemlets branching from the main leaf stem, each bearing 15 to 25 pairs of oval leaflets. The leaflets are ¾ of an inch long by ¼ inch wide. In late Spring and early Summer, clusters of greenish-white flowers, each a tuft of ½ inch-long stamens, appear on the tree, and are followed by flattened, papery pods with seeds oriented across the width of the pod in a row down its center. These trees come from the Moluccas, New Guinea, New Britain, and the Solomon Islands. The tree has been widely planted for reforestation and is now naturalized, especially in valley bottoms, on all the major Islands. Reportedly, it was introduced to Hawai'i by Joseph Rock as an ornamental and for reforestation in 1917.

Like many legumes, paraserianthes forms a nitrogen-fixing association with bacteria. The tree is probably most easily recognized in the distance by its wide-spreading flat-topped crown which can often be seen from ridge trails in the valleys below. The people who built the ditches to bring water from the mountains to the sugar cane fields sometimes planted paraserianthes along the ditch trails, and very large trees can still be found along these paths. This tree grows very rapidly, but is not very strong, and wind storms often topple whole trees or break off major limbs, tearing out or obstructing sections of our trails. Timber from this tree has been used for corestock veneer, pallets, boxes, and shelving. It may be a useful pulpwood for making paper.

ROSE APPLE
Syzygium jambos
Myrtaceae (manuka, ʻōhiʻa, guava, eucalyptus)

These are trees with many branches that are 20 to 40 feet tall with spindle-shaped opposite leaves that are about 4 to 10 inches long by 1 or 2 inches wide. The leaves are dark green above, a lighter dull green beneath with a noticeable vein running near each margin. Cream colored puff-ball flowers made up of filaments (stamens) up to 2 inches long are seen in season and are followed by round yellow edible fruit 1 or 2 inches in diameter. Each fruit bears 4 prominent lumps on one end which are the remains of part of the flower. The fruit is crisp but

Look for: Trees with long, thin, spindle-shaped opposite leaves that are dark green in color. In season, observe white to cream powder-puff flowers or round yellow rose-flavored fruit.

insipid with little juice and a large cavity that usually contains a single seed, and has the flavor and smell of rose perfume. The original home of this tree is uncertain, but it probably comes from south east Asia or the adjacent islands. In Hawaiʻi it grows wild in low elevation mesic and wet forest, especially in valleys. Well established groves cast such dense shade that nothing else, except rose apple seedlings, can grow under them. The tree is thought to have been introduced to Hawaiʻi for its edible fruit in 1825 from Rio de Janeiro by the frigate *Blonde*. This ship, under Captain Byron, cousin of the famous poet, was sailing to Hawaiʻi to return the bodies of King Liholiho and Queen Kamamalu who had died of measles during a royal tour of England. Recently a fungus (perhaps *Puccinia psidii*) that attacks the new leaves on this and related native plants (nīoi, ʻōhiʻa, ʻōhiʻa hā) has been found in the Islands and appears to be devastating to the trees. The future of this plant is therefore in some doubt, and the 2 endemic species of nīoi, already uncommon, may well become extinct.

SUGI, TSUGI
Cryptomeria japonica
Taxodiaceae (redwood, sequoia)

Sugi is the national tree of Japan. It was introduced into Hawai'i in the 1800s and has been extensively used for forestry plantings, some 500,000 seedlings being set out between 1920 and 1960 alone. This is an evergreen tree that may grow to nearly 200 feet in favorable conditions with a straight trunk that tapers from an enlarged base. The bark is reddish-brown with long fissures and is slightly shaggy and fibrous, tending to peel off in long shreds. The branches spread horizontally and bear densely clustered sharp needle-like leaves about ½ inch long that curve forward. Seeds are produced in reddish cones about 1 inch in diameter, consisting of thick scales, each with 4 or 5 coarse teeth. Sugi is seen along trails, often in sheltered gulches and at moist middle elevations on all the larger Islands.

Look for: An evergreen tree with short sharp densely clustered needles on a tree with reddish-brown somewhat shaggy bark.

Although often called "sugi pine" this tree belongs to a somewhat more primitive family of gymnosperms than the pines, a family that includes the California redwood and sequoia as indicated above. Most of the forest in Japan is made up of sugi, although this may not be its original state but may represent many generations of forestry plantings by humans. If the timber of this tree is buried in the ground it will turn dark green, producing *jindai-sugi*, which is considered a kind of semi-precious "stone".

TOONA, AUSTRALIAN RED CEDAR
Toona ciliata
Meliaceae (mahogany, pride of India, neem tree)

This is a tree up to 50 or more feet tall with large alternate pinnate compound leaves about 1 or 2 feet long with 10 to 14 leaflets, each a narrow, lance-shaped blade 2 to 7 inches long. The terminal leaflet(s) may be single or paired. Large clusters of small white flowers appear in season near the ends of the twigs. These are followed by oval brown seed capsules about ¾ of an

KEN SUZUKI

140

inch long, that open by means of 5 valves. The trees have light gray bark. The tree is native to India, south east Asia, and eastern Australia. This species was apparently introduced into Hawai'i about 1914 and has been used for reforestation in moist areas. It is now found in such habitats on all the larger Islands and is considered an undesirable invasive weed. Young trees are similar to tropical ash, but have alternate rather than opposite leaves.

Look for: A gray-barked tree of mesic to wet forest with large alternate pinnate compound leaves with long lance-shaped, often curved, leaflets.

TROPICAL ASH
Fraxinus uhdei
Oleaceae (olive, olopua, pīkake, jasmine)

These trees may reach more than 70 feet in height, with gray or brown furrowed bark. The leaves are opposite and odd-pinnately compound with 5 to 9 leaflets, each a pointed oval 2 to 4 inches long, sometimes with small irregular teeth along each edge. The flowers are inconspicuous but the fruit is a typical "key" with a small oval

Look for: A tree of mesic forest with opposite odd-pinnately compound leaves and inch-long single- winged fruit.

body about ¼ inch long at one end of a flat oblong wing about an inch in length. The trees are native to central and southern Mexico and were introduced to Hawai'i for reforestation. Between 1924 and 1960 about 700,000 trees were planted on all the larger Islands. They now appear to be actively reseeding themselves in mesic forest and spreading.

Saplings of this tree differ from those of the African tulip tree in that the leaflets of the latter tend to be broader, have sunken veins, and never have teeth. Toona can be distinguished from both in having alternate leaves. The seeds of each are also quite different.

TURPENTINE TREE
Syncarpia glomulifera
Myrtaceae (Java plum, eucalyptus, guava, 'ōhi'a hā)

This is a large tree up to 100 feet tall which is closely related to eucalyptus. It differs from the latter in having a number of small flowers, and later seed capsules, fused into a ball. The tree has gray to reddish-brown bark that is thick and deeply furrowed, fibrous and shaggy. The leaves of this tree are lance-shaped blades 2 ½ to 5 inches long by 1 to 1 ¾ wide. The 2 to 4 leaves at a node are a dull green above and a dull whitish green beneath. Flowers appear in a small white puff-ball which is made up of 6 to 11 individual blooms joined at the base. The fruit consists of 6 to 11 hard woody capsules fused at the base to form a disk with crown-shaped openings about ¼ inch across around the rim, each representing an individual capsule. One or more additional capsules may protrude from the center of this disk. The wood of this tree is resistant to decay and to termites and is difficult to burn. The bark yields turpentine. This Australian native has been widely used for reforestation in Hawaiʻi, especially on Oʻahu. About 83,000 trees have been planted in the Islands and may be seen along a number of our trails.

Look for: A large tree with lance-shaped leaves, dull green above and whitish green below, 2 to 4 to a node, and hard woody capsules fused to form a cluster with 6 or more round, crown-shaped openings around it.

ʻULU, BREADFRUIT
Artocarpus altilis
Moraceae (fig, banyan, wauke)

These trees were introduced to Hawaiʻi by Polynesian settlers in prehistoric times. They were probably a relatively late arrival, however, since only a single form was found here, in contrast to the numerous local varieties of taro and sweet potatoes that were developed in the Islands. Many other varieties of breadfruit occur in islands further south, and some of these have been brought to

Look for: Trees of low elevation valleys with huge leathery leaves divided into prominent lobes and large round warty fruit.

Hawai'i in more recent years. 'Ulu trees can grow to be 60 feet tall and are distinguished by the very large hairy alternate leaves, 8 to 20 inches long by 6 or 8 inches wide, with deep indentations along the sides that divide them into large blunt lobes. In season, the trees bear large oval or round starch-rich warty fruits, 6 inches or more in diameter, that can be cooked in a variety of ways. These were a major staple of other Polynesians, in the Marquesas Islands in particular, but a minor item in the diet of the Hawaiians. The Hawaiian variety lacks seeds. The tree probably came from Micronesia originally, but has been cultivated as an important food plant since ancient times and is widespread throughout Pacific cultures. You are most likely to see it along lowland valley trails, where it marks the sites of former homesteads.

Besides their value as a source of food, 'ulu trees were hollowed out to make small canoes, or drums, or carved into poi pounding boards and small surf boards. The bark can be beaten into kapa cloth, although the Hawaiians much preferred that obtained from the wauke, or paper mulberry, for this purpose. The sap could be used to caulk canoes, as a bird lime to trap small forest birds for their feathers, or as chewing gum. It was also a component in a mixture to treat infected sores or applied to cuts and cracked or chapped skin. A tan or yellow dye was obtained from the male flower, and the sheath shed from newly emerging leaves made a fine grade of sandpaper. One legend has it that the spirits of the dead leaped into the underworld from the branches of a large 'ulu that grew near the lakes in Moanalua Gardens in Honolulu.

VINES

ʻĀWIKIWIKI
Canavalia spp.
Fabaceae (tamarind, lupine, chick pea, beans)

KEN SUZUKI

Look for: A tree-climbing vine with typical bean-like leaves and sweet pea flowers in shades of purple or red.

There are 8 species in this genus in Hawaiʻi, of which 6 are endemic and 2 are introduced. They are all vines that climb into shrubs and trees, with compound leaves that have 3, bean-like, large pointed oval leaflets. The plants produce small spikes of sweet pea-like flowers that range in color from pink to various shades of red and purple to magenta in different species. In the Oʻahu species, *C. galeata*, which has dark purple flowers, each leaflet is about 2 to 8 inches long by 1 to 3 inches wide. Narrow pods about 5 inches long and an inch wide contain flat brown seeds that are sometimes mottled with black. The vines are found scattered in mesic forest, often in areas that have been invaded by guava or even lantana. Each of the native species tends to be confined to 1 or 2 Islands although several Islands have more than one.

ʻĀwikiwiki demonstrates the harm that feral animals have done to our native flora. In the 1960s, a few small areas of Volcanoes National Park on Hawaiʻi were fenced to exclude goats in an effort to determine what impact these animals had on the Park vegetation. In one of the lowland dry exclosures, an unknown species of *Canavalia* began to grow and dominate the area. Apparently seeds of this species had lain dormant in the soil for 120 years or more, with a few sprouting every time weather conditions were favorable, only to be devoured by goats or pigs before they could mature. With the animals excluded, they could once more assume their natural place in the ecosystem. How many more of our lovely native plants have never been seen because of the depredations of these animals? How many have vanished forever because they did not produce such long-lived seeds?

Early Hawaiians used vines of these plants for the frames of scoop nets used to capture small fish and shrimp. They were also woven into rude and temporary fish traps. The flowers were used in lei, and the whole plant was pounded into a mash with other plants to produce a medicinal fluid used to treat itchy skin and other skin disorders.

CAT'S CLAW, WAIT-A-BIT
Caesalpinia spp.
Fabaceae (golden shower tree, clover, carob)

The plant in this genus that you are most likely to encounter is *C. deca-petala*. This is called "cat's claw" or "wait-a-bit" and is a really nasty customer to find on a trail as it forms dense tangles of sprawling thick branches that are covered with cruel sharp thorns that are found even on the leaves. It has doubly compound leaves with the main leaf stem bearing 3 to 15 pairs of stemlets each of which has 5 to 12 pairs of oval leaflets up to 1 inch long and ½ inch wide. Large erect clusters of lovely yellow flowers are followed by 2 to 4 inch long bean pods. This plant is native to tropical Asia and was introduced into Hawai'i as a hedge plant for ranches, since no cattle are likely to try to force their way through it. It was first collected on O'ahu in 1910 and is now common in lowland valleys and on adjacent slopes near former ranches on all the major Islands.

Look for: A really evil plant forming dense thickets with its tangle of sturdy stems, that has attractive green compound leaves and clusters of lovely yellow blossoms, but is covered with cruel and abundant thorns.

Two other species in this genus, *C. bonduc* and *C. major*, are viny shrubs with doubly compound leaves and small hooked thorns on the stems and leaf stalks. Both are called "kākalaioa", meaning "prickly" in Hawaiian. There is some uncertainty about whether they are indigenous natives or early introductions to the Islands, but both are found scattered around the tropical world in dry forests near the sea coast. Each leaf has 3 to 9 pairs of stemlets branching from the main stem, with 3 to 9 pairs of leaflets arranged along each of these. The leaflets in *C. bonduc* are oval and about ½ inch to 2 inches long. Those of *C. major* are larger, up to 5 inches long by 2 wide. Clusters of small yellow flowers are followed by large, prickly bean pods. The plants are found on all the major Islands. *C. bonduc* was also known as "gray nickers" because the seeds are pale gray to olive gray. *C. major* was called "yellow nickers" because the seeds, when fully ripe, were grayish yellow. So if you find the plant with mature seeds on it, you should be able to tell which one it is. I am

145

only familiar with *C. bonduc*, which is found in the Waiʻanae Mountains on Oʻahu in the Nature Conservancy Honouliuli Reserve, and this is the one I describe below.

The seeds of gray nickers, and perhaps yellow nickers as well, were used as marbles by children and could also be strung as beads for lei. They have also been used medicinally, especially in India. The powdered seed is a strong purgative.

Look for: A leggy, stiff, viny shrub with few branches and widely spaced compound leaves with a few small curved thorns on the stem and leaf stalks, in mesic lowland forests.

HOI KUAHIWI, SMILAX
Smilax melastomifolia
Smilacaceae (greenbrier, catbriar)

About 300 species of *Smilax* occur around the world in both tropical and temperate regions. This one is endemic to Hawaiʻi where it is found in upper mesic and wet forests. The thick heart-shaped leaves are glossy and have prominent veins running from base to tip, much like a melastome, although it is not related to that family. As the Hawaiian name implies, its 4 to 6 inch long

KEN SUZUKI

Look for: A vine in upper mesic and wet forests with glossy, heart-shaped leaves that clings to other plants by means of tendrils, and has prominent veins running from base to tip of the leaf.

heart-shaped leaves resemble those of one of the yams that the Polynesians introduced here—hoi kuahiwi means "mountain yam", which is appropriate since it is found scattered along ridges in the mountains and not just in the shady gulches where hoi itself is found. The plant is a vine and a pair of tendrils emerge from the stem near the base of the leaf and help it cling to the other plants that it clambers over. Thus it is one of our few native lianas or viny plants. The flower is small, about ½ inch in diameter, with 6 petals, rather like a small lily, and greenish white to white in color. This is followed by a green or blue berry about ⅓ inch in diameter. Male and female sexes are on different plants. The stems may have prickles, but many plants lack these, illustrating the tendency of Hawaiian plants to lose their protection against large herbivores due to the lack of such animals in the Islands in the past. Relatives in the mainland United States have this armament as the common names above indicate. The tendrils and prickles help distinguish this plant from hoi, which lacks these features.

The plant has a starchy tuberous rhizome or underground stem, that may have been used as a famine food by the Hawaiians, but apparently was not very palatable. A tropical American species has roots that produce the flavoring material, sarsaparilla.

'IE'IE
Freycinetia arborea
Pandanaceae (hala or pandanus)

This indigenous native climber is also found in New Caledonia and many other islands of the South Pacific. In Hawai'i it is common in the upper mesic and wet forest. It resembles hala pepe, but unlike that tree, it does not form a trunk and so is not able to stand erect for any distance by itself, but sprawls on the ground, forming dense tangles, or uses its aerial roots to cling to trees and make its way up into the crown. Unlike most tropical rain forests, Hawai'i has very few native vines or lianas, but 'ie'ie is one. The stems of the plant are an inch or more in diameter and have rings around them where the scars from old leaves have left their mark. They branch every few feet and the stems end in tufts of narrow, spiny leaves that may be up to 2 ½ feet long. These have prominent mid-ribs and clasp the stem at one end while they taper to a slen-

der point at the other. These leaves often have small sharp teeth on the mid-rib and leaf edges near the end of the leaf, which distinguishes them from the smooth leaves of hala pepe. The flower consists of 2 to 4 cylindrical bodies that rise from the end of the stem and are surrounded by broad, edible, orange-colored modified leaves. These can be quite attractive when well developed, although the flowers are usually spoiled by rats which eat them. Many long, narrow aerial roots emerge along the stem and help the plant hold on to the trees it grows on. Some of these may be as long as 20 feet, and if they reach the ground, will become rooted in the soil.

JOHN HOOVER

Look for: A thick stemmed vine that climbs into the trees, with tufts of long slender pointed leaves at the end of the stem and long aerial roots.

The Hawaiians had many uses for the long, pliable but tough aerial roots of 'ie'ie. These could be used whole to make rough fish traps, or steamed in the imu and split to make finer traps, baskets, feathered helmets for warrior chiefs, or the feathered gods that were carried into battle. In the feather work, the feathers were usually attached to fine nets of olonā fiber which were then sewn onto the underlying basketry. Hawaiian baskets were often skillfully and beautifully woven and were probably the finest in Polynesia. The roots could also be used simply as cordage.

According to legend, as recounted by Marie C. Neal, in *In Gardens of Hawaii*, Laukaieie (leaf of the 'ie'ie) was a child who was carred for by the goddess Hina. She was given to a lonely couple in the forest to raise, and there she had no playmates but the flowers and the birds. Laukaieie grew to be a beautiful maiden and she married a bird-man. For reasons that are not explained, the time came for her to change form—she began to sprout leaves on her slender body and her eyes glowed. Her husband took her to the forest and said, "You can not stand alone. Climb trees! Twine your long leaves around them. Let your blazing red flowers shine between the leaves like eyes of fire. Give your beauty to the all the 'ōhi'a trees of the forest!" And so the maiden became the 'ie'ie vine, and a flowering branch of this plant is placed on the altar in the hālau hula to represent the demi-goddess Laukaieie.

LILIKO'I, PASSION FRUIT
Passiflora spp.
Passifloraceae (I know of no familiar relatives)

At least 12 of the many species in this genus have been introduced to Hawai'i, all from the American tropics. Several of these produce delicious edible fruit, but at least 3 are major pests and are a threat, over large areas, to the survival of our native ecosystems. All of these plants are vines which climb on other plants by means of tendrils. Their leaves are usually alternate with 3 lobes, and they have complex flowers of a unique form. Most bear fruit with a hard or leathery rind and numerous seeds surrounded by a slippery, juicy edible pulp. I will briefly describe the species that you are most likely to encounter.

P. edulis has leaves with 3 prominent lobes that are rounded or bluntly point-ed. The leaves are 3 to 6 inches long and about the same width. The flow-ers have 10 white petals, purple centers and numerous filaments that radiate from the center. A cluster of prominent foot-shaped or golf-club-like repro-ductive structures arise from the flower center. The oval yellow fruit is about 1 ½ inches in diameter and contains a tart pulp that is used to make passion fruit juice, jellies, etc. It is quite good eating, but as for all liliko'i, should not be chewed, but just mashed a bit with the tongue and swallowed, as the slippery pulp allows the seeds to slide down easily. A variety of this species has somewhat smaller purple fruit with a sweet pulp that I used to love. It was once common on Oʻahu and especially in the Kōkeʻe area of Kauaʻi, but is scarcely to be found on the former Island now, and seems less common on the latter also. This plant is found scattered in mesic forest on all the major Islands. It is not uncommon, but rarely becomes a pest. It arrived in Hawaiʻi sometime before 1871, when it was found growing on Maui.

P. laurifolia, yellow water lemon, differs from most of its relatives in having unlobed oblong leaves 2 to 6 inches long. The flowers are very similar to those of the preceding species, but the leathery fruit is orange and oblong, about 2 inches long, and with discernable grooves running from tip to stem that divide it into 3 or 4 segments. The pulp has a sweet, slightly soapy flavor that I enjoy, but it is not relished by everyone. It was apparently cultivated in Hawaiʻi before 1871 and is now a major pest in some areas. On the lowlands of windward Oʻahu it has spread over large tracts, smothering virtually everything else in the area.

KEN SUZUKI

KEN SUZUKI

Look for: A tree-climbing vine of mesic forest with deeply lobed, 3-fingered leaves and 1 ½ inch diameter yellow fruit, or typical passion fruit flowers.

Look for: A spreading vine of lowland mesic areas with unlobed oblong leaves and oblong orange fruit with a sweet, slightly soapy-tasting pulp.

P. foetida is a dryland lilikoʻi with 3-lobed pointed leaves and a small red, marble-sized berry enclosed in a lacy basket. You can not mistake it when you see it in fruit. The fruit may be edible, but probably not worth the trouble.

P. mollissima or banana poka is a really nasty customer. This malicious plant was introduced as an ornamental, rather than for its fruit—supposedly to screen an outhouse on the Big Island, where it was first collected in 1926. It has the typical deeply lobed 3-fingered leaves of many lilikoʻi, but the flower is a large pink beauty in which 10 petals emerge at right angles to a long tube with a bulbous base. A 2 inch long rod extends from the center of the flower and bears the reproductive structures. The fruit is an elongated yellow oval "banana" about 3 inches long, containing a pulp which some people eat, although I do not find it palatable. Pigs feed on it, however, and help to spread the seeds. This plant is a major pest on Kaua'i, Maui, and the Big Island where it clambers into the tallest trees and smothers them in large areas of moist native forest. An effort is being made to introduce insects and diseases that might weaken it. Fortunately it is not yet established on O'ahu, although some people mistakenly call *P. laurifolia* "banana poka".

VINCENT T. SOEDA

Look for: A high climbing vine of moist forest with striking pink 3 inch diameter blossoms and elongated yellow fruits somewhat like small bananas.

P. suberosa or corky passion vine is a member of the family with no redeeming qualities. It must have been introduced by accident, since it has neither edible fruit nor attractive flowers and has become a really serious pest. This is a vine of the dry forest with 3 lobed, pointed leaves that, when young, can be mistaken for those of the native huehue. The plant is easily recognized however by the sheath of deeply furrowed spongy, corky material that encloses the vine in older plants. The flower is less than an inch in diameter, greenish, and of typical passion fruit structure. It is followed by small oval black berries about ½ inch in diameter that only a bird could love. This vine is a menace in dry and mesic forest where it smothers all other plants over wide areas. The Gulf fritillary, *Agraulis vanillae*, whose larvae feed on passion fruit plants arrived here by some means, probably from California, about 1971, and is now often seen wherever these plants grow. The caterpillars have had very little impact on the health of the infestation as far as I can see, however. This black and orange butterfly

KEN SUZUKI

Thomas H. Rau

resembles the Kamehameha butterfly, one of only 2 native butterflies in the Islands.

Look for: A tangle of vines sheathed in deeply furrowed spongy material and having 3-lobed, pointed leaves in dry and mesic forest.

I have read that a group of teenagers belonging to a strict fundamentalist Christian sect developed a great thirst for passion fruit juice, this being the most risque behavior that they were allowed. Actually, the name "passion fruit" had nothing to do with sex, but was conferred on the plants by early Spanish colonists, possibly Jesuits, in the Americas, who saw in the unusual structure of the flower an elaborate set of symbols of the Passion of Christ. The 10 petals represented the 10 apostles present at the crucifixion, the crown of filaments was the crown of thorns or perhaps a halo, the 5 stamens were the 5 wounds of Christ, the 3 styles were the 3 nails, the tendrils were the cords that bound Him or perhaps the whips that scourged Him. The white petals symbolized purity and the blue center, heaven. The flower was supposed to last for 3 days, standing for the 3 years of His ministry.

MAILE
Alyxia oliviformis
Apocynaceae (periwinkles, plumeria, oleander, allamanda, the be-still tree)

Maile is an endemic viny shrub that often forms dense tangles of branches in favorable locations. It may be found from the upper dry forest well into the wet forest. The shiny dark green oval pointed leaves are opposite in pairs or sometimes threes or fours. They are highly variable in size, from 1 to 4 inches long, or more. Maile resembles one of the pilos, *Coprosma folliosa*, but its leaves are a darker green and glossy, and if you pluck one of the leaves, a small milky drop will form at the break. As the species name implies, the fruit is a small black olive-like berry. The flower is tiny—about ⅛ of an inch across, with 4 small yellowish petals on a longer tube that has an orangish tint.

JOHN HOOVER

Look for: A vine or viny shrub with small oval pointed dark green opposite leaves that have a milky sap.

All parts of the plant contain the chemical, coumarin. When the bark and leaves are carefully stripped from a length of branch, they develop a pleasant fragrance due to this substance. Maile was one of the 5 essential plants that were placed on the altar of Laka, the goddess of hula. The plant represented the 4 Maile sisters, legendary sponsors of hula. Lei made of maile were also highly valued, and there is still a great demand for these lei, especially at graduation time. Maile was one of the plants placed in calabashes where kapa was stored in order to impart some of its perfume to the kapa cloth.

MAILE PILAU
Paederia foetida
Rubiaceae (coffee, gardenia, cinchona, manono)

The Hawaiian word "pilau" means a nasty smell as does the Latin "foetida". My dull nose does not find anything terribly unpleasant about the odor of this plant, but who am I to argue with the experts? You can judge for yourself. The plant is a twining, slender, bad-smelling vine 6 to 20 feet long with well-spaced opposite leaves. The leaves are pointed ovals about 1 ½ to 5 inches long by an inch or two wide. Large open clusters of flowers emerge from the tips of the vine or from the junction of leaf and stem. They are rather at-

Look for: A slender, twining, stinky vine with opposite leaves well spaced along the stem and clusters of small attractive bell-like flowers with maroon centers surrounded by much paler, somewhat ragged petals.

tractive pale lavender bells about ½ inch long that open into 5-lobed, ragged-edged blossoms, about ½ inch wide, with deep maroon throats. The oval fruit are yellowish-brown to red and about ¼ inch in diameter. The plant is native to east Asia and is now common in coastal, dry, and mesic forests as well as in subalpine woodlands in Hawai'i. It is not known how it got here, possibly being introduced as an ornamental, and was first reported on O'ahu in 1854.

The seeds of this plant are spread by birds and it has become a serious pest in some areas, weighing down overhead wires and fences and smothering desirable trees and shrubs.

NUKU 'I'IWI
Strongylodon ruber
Fabaceae (beans, shower trees, acacia)

This plant is an endemic vine with bean-like compound leaves displaying 3 broadly oval pointed leaflets, each about 3 or 4 inches long. Flowers are borne on long hanging pendants and are bright scarlet. Each blossom is 1 to 2 inches long with a slender curved beak that resembles the bill of the native 'i'iwi bird.

Look for: A vine with 3-part, bean-like compound leaves and lovely sharp-pointed scarlet flowers high in the trees in mesic to wet forest.

("Nuku" means mouth or bill). The pods are broad ovals about 3 or 4 inches long, often with a sharp hooked point at the end. The vine can be found spreading through the trees in mesic to wet forests on all of the higher Islands.

The flowers were sometimes used in lei. The jade vine with its curiously colored blue-green flowers that is sometimes seen in gardens in Hawai'i is a member of the same genus. This vine comes from the Philippines and its flowers are similar in form to nuku 'i'iwi but considerably larger. When in blossom or bearing pods, this plant is easily distinguished from 'awikiwiki, but when they are not in flower, which is most of the time, I know of no easy way to tell them apart.

Thomas H. Rau

FERNS

ʻAMAʻU
Sadleria spp.
Blechnaceae (blechnum)

This is an endemic genus of 6 species. Four of these are medium to fairly large tree ferns found in a variety of habitats from recent lava flows to mesic and wet forest, while the remaining two are small ferns of dark, wet stream banks. The species that you are most likely to encounter is *S. cyatheoides* although the other 3 large tree ferns are

JOHN HOOVER

Look for: A palm-shaped tree fern with feather-like fronds that are salmon colored when young.

very similar to it. It has a short trunk, usually 2 or 3 feet tall, although in favorable locations it may grow to 9 feet or even more. The leaf stalks cluster at the top of the trunk like the fronds of a palm, are straw-colored and have brown hairs covering the lower parts. The leaves are 3 or 4 feet long, doubly compound, with rows of small stalklets on either side of the stalk and 30 to 60 pairs of small oval leaflets attached to each of these. The overall form is that of a large coarse feather. Young fronds are a bright salmon color. The sori, or spore-bearing bodies, are found on the underside of a leaflet and are bar-shaped. This fern is one of the first plants to colonize a recent lava flow on the Big Island and is also found in dry, mid-elevation shrubland, open woodlands, rain forest, and subalpine zones.

ʻAmaʻu is a kino lau or plant form that the pig-man demigod Kamapuaʻa can assume at will. A red dye for staining kapa cloth was extracted from the reddish young fronds, or perhaps from the cortex of the trunk. The young fronds could be cooked and eaten, as could the starchy pith, though this was probably only done during times when preferred food was lacking. The raw pith could be fed to pigs. It was also a component of a medicine to treat asthma. The mucilaginous sap from the leaf stalk was used to bind pieces of kapa together. Fronds were sometimes used to mulch fields of dryland taro or to decorate the ridge pole of a house. The mass of fine hairs, or pulu, from the base of the leaf stalks was used to stuff pillows (though not, perhaps, until modern times) and in embalming the dead. The main fire pit in the caldera of Kīlauea is named "Halemaʻumaʻu" or House of the ʻAmaʻu Fern". These ferns, unfortunately, are often attacked by wild pigs that knock them over and eat the starchy pith in the trunk and kill the plant. This also leaves a hollow in which water collects and mosquitoes can breed, which in turn is a disaster for our native birds which are often killed by the bird pox and bird malaria transmitted by these insects.

AUSTRALIAN TREE FERN
Sphaeropteris (Cyathea) cooperi
Cyatheaceae (no familiar relatives)

These are tall, slender tree ferns that have become very popular as landscape plants. They have erect trunks that may reach 30 feet or more in height but are much more slender in diameter than our native hāpuʻu. On older ferns, the oval scars left by fallen fronds are conspicuous. The circle of fronds at the top of the trunk forms an attractive lacy, funnel-shaped crown. At the base of the frond stalks a mass of light tan material like finely shredded paper caps the top of the trunk and surrounds the emerging young fronds. Fronds may be 10 feet long and 3 feet wide. Each frond is divided 3 times, as in hāpuʻu, and the spore-bearing sori occur on the lower surface of the leaf as a row of small brown round bodies on either side of the midrib of a lobe. This fern was introduced into Hawaiʻi about 1940 as an ornamental. It is native to northeastern Australia. It is hardy, fast growing, and tolerates drier conditions and more sun than most of our native tree ferns. Unlike them, it does not seem to be attacked by pigs. As a result, it is now aggressively spreading in mesic and wet forest on all the major Islands.

Look for: A slender tree fern with a lacy, funnel-shaped crown and a mass of light tan material like finely shredded paper at the base of the fronds.

Our native hāpuʻu has a thicker, fibrous trunk and fine hairlike material around the base of the frond stalks. This hair is golden to reddish to dark brown in color, but not light tan. Unlike the native fern, the trunk of mature Australian tree ferns is not enclosed by a fibrous root mass except at the base, although in young ones this may not be apparent.

BIRD'S NEST FERN, 'ĒKAHA
Asplenium nidus
Aspleniaceae (no familiar relatives)

This large fern is usually seen perched on the limb of a sizeable tree, often a kukui, but may be found on the ground. The fronds are a simple spindle-shape without branches or lobes and are 2 to 4 feet long by 3 to 8 inches in width. They are arranged around the base of the fern like the feathers of a badminton shuttlecock, forming a broadly conical nest-like structure. The midribs of the fronds are dark and prominent. When fertile, the dark colored spore-bearing sori are seen as thin, closely packed lines on the upper ½ to ⅔ of the underside of the frond, and running from the midrib to about ⅔ of the way to the margin. These ferns are relatively common in trees in low elevation moist gulches on all the main Islands. This is an indigenous fern, being found also throughout Polynesia and in tropical Asia and Australia, and west to Madagascar.

KEN SUZUKI

Each frond resembles a giant stag's tongue fern, which the Hawaiians also called "'ēkaha". They used the dark midribs of these ferns to make patterns of a contrasting color in their lau hala mats. After cutting down a large tree to make a canoe, they would carry out a ceremony in which 'ēkaha was used to cover the remaining stump. The fronds were also used to decorate the hālau hula.

Look for: A large fern perched on tree limbs in gulches in mesic and wet forest and looking like a conical basket or nest.

BLECHNUM
Blechnum appendiculatum
Blechnaceae ('ama'u)

This is a medium sized fern with pinnately compound fronds about 8 to 24 inches long and the overall shape of a dagger. Two rows of leaflets line the central stalk. These have a rounded, somewhat bulbous base and taper with a slight curve to a blunt point. Young fronds are conspicuous for their salmon color. The spore-bearing structures are thin lines that closely flank

the midrib of the leaflet. The fern forms large colonies in shady mesic forest on all the major Islands. This introduced plant is native to tropical America and was first collected in Hawai'i in 1918. It is sometimes grown in gardens and was probably introduced as an ornamental. It has now become a weed in our forests where it crowds out native plants.

Look for: A medium sized, pinnately compound fern with an overall dagger shape. It is easily recognized when young fronds are present by their bright salmon color.

BRACKEN FERN, KĪLAU
Pteridium aquilinum
Dennstaedtiaceae (palapalai)

These ferns are medium-sized plants growing on the ground, with widely spaced fronds. The fronds are divided 3 times, the overall form being triangular and lacy, and the branches well separated, giving the frond a fairly open appearance. The plant may be 6 inches to 3 feet tall, depending on habitat. The stalks are shiny, hairless, and yellow. The leaflets near the base are highly lobed while those in the middle of the frond are simple oblong blades and near the tip they are no longer separate but are reduced to being lobes of a single leaflet on the secondary stem. The spore-bearing sori, when the frond is fertile, form a continuous line near the margin of a leaflet. This is a common fern of mesic and wet forest and also occurs in dry subalpine regions above 8,000 feet in altitude.

This fern is one of the few plants in Hawai'i that dies back in the winter. It competes well with alien grasses in dry woodlands, but is subject to attack by pigs, who grub up the underground stems and eat them. There is some controversy about the classification of these ferns. If all bracken ferns are indeed members of one species, they are among the world's most successful plants, being found from the alpine deserts of Haleakala to rainforests of northwestern America, and indeed, essentially

KEN SUZUKI

Look for: A lacy, open, well-spaced terrestrial fern with triangular fronds generally found in fairly sunny, open habitats.

everywhere in the world except Antarctica. The variety, *P. aquilinum decompositum*, is considered endemic to Hawai'i, however.

157

CHRISTELLA, DOWNY WOOD FERN
Christella parasitica, C. dentata
Thelypteridaceae (no familiar relatives)

There are 5 species in this genus in Hawai'i plus several hybrids between two species. Three of the 5 are endemic, but the 2 above, which are by far the most common, are alien introductions. These 2 ferns are erect with unbranched stems along which 2 rows of deeply lobed leaflets are arranged in the pinnately compound pattern. They are generally 1 to 3 feet tall. The overall form is blade-like with the widest part being near the middle and tapering to a sharp point. The fronds are usually broader and

Look for: A very common soft fuzzy fern whose pinnately compound fronds have a broad dagger-shape made up of a single stem bearing 2 rows of long, deeply lobed, pointed leaflets.

shorter than in the sword ferns and the leaflets are longer and more sharply pointed than in the latter. The fronds are soft and fuzzy. The spore-bearing sori form small brown dots that lie in rows on either side of the midrib of a lobe of the leaflet. *C. parasitica* is native to Australia, southeast Asia, Japan, and adjacent islands. It was first reported in Hawai'i in 1926 and is now probably the most common fern in this genus in the Islands. *C. dentata* is widely distributed in the Old World tropics and has become a wide-spread introduced weed in the Americas. It was first collected in Hawai'i in 1887 and soon became a commonly encountered fern here. Both species are now found in mesic to wet forest on all the Islands. The differences between these 2 are very minor in the eyes of the nonspecialist.

'ĒKAHA, HART'S (or STAG'S) TONGUE FERN
Elaphoglossum spp.
Lomariopsidaceae (no familiar relatives)

There are 8 endemic and one indigenous species in this genus in Hawai'i, plus several hybrids. These are small to medium ferns that grow on the ground or on trees. They are the simplest possible ferns in shape, having spindle-shaped unlobed fronds on an unbranched stalk.

They are generally about a foot tall or less, and 1 or 2 inches in width. One endemic species, *E. crassifolium*, has a fine network of veins. It is probably our most common 'ēkaha. All the other species have parallel veins. Another, *E. paleaceum*, is covered with a fur of brown hairs. The spores are born on the undersurface of fronds that are specialized for this purpose, and usually differ in form from the sterile fronds. These ferns are found in upper mesic to wet forest on all the main Islands.

Look for: Simple, leathery, spindle-shaped ferns growing in mesic to wet forest on the ground or on trees.

HĀPU'U
Cibotium spp.
Dicksoniaceae (no familiar relatives)

There are 4 endemic species of this genus in Hawai'i. The trunks of the largest may grow to be 20 feet in height and nearly 2 feet in diameter at the base. The fronds arise, palm-like, from the top of the trunk, and may be 20 feet long. The ones we see on O'ahu are considerably smaller, however. The frond stalks are hairy at the base, and in some species, along much of their length. They do not have

KEN SUZUKI

Look for: A large tree fern with a fairly thick trunk, a mass of golden to black hairs at the base of the fronds, and lacy triangle-shaped fronds in upper mesic and wet forest.

the pale, papery material characteristic of the Australian tree fern at the base of the fronds. The fronds are triply compound, with stalklets branching from the main stalk, and then sub-stalklets branching from these, with the deeply lobed leaves arranged along both sides of the last division. The overall form is open and lacy and roughly triangular in shape. The spore-bearing structures, or sori, are cup-like and appear near the margins of a leaflet. While these ferns are found in mesic forest on all the Islands, they reach their maximum development in wet forest, such as is found along the Crater Rim drive in Hawai'i Volcanoes National Park on the Big Island.

The fiddle heads, or uncoiling young fronds, could be cooked and eaten, and the starchy pith in the center of the trunk was used as a famine food, or fed to pigs. In more recent years, around the 1920s, this starch was harvested and sold for use in cooking or the laundry. The mass of fine hairs, or pulu, found at the top of the trunk was used as a wound dressing by early Hawaiians and in embalming. Between 1854 and 1884, several hundred thousand pounds of this pulu was harvested on the Big Island and shipped to the U.S. mainland to use in stuffing pillows and mattresses. Fortunately for our native forests, it did not prove to be satisfactory in the long run, as it soon disintegrates into a sandy grit. Some hāpuʻu is still harvested however for the fibrous outer trunk, actually a mass of roots that surround the true stem of the plant, which is used as a medium for growing orchids and other plants. The Hawaiians built bins of hāpuʻu logs in which to plant yams. They also used the pulu as an ad-sorbent for several liquid medicinal mixtures to be inhaled or inserted in the nose to treat nasal growths, stuffy nostrils, or sinus congestion. The pith was a component in a complex concoction intended to "clean the blood". Like the ʻamaʻu fern, hāpuʻu is subject to damage by pigs which knock over the trees to eat the starchy pith, leaving hollows that fill with water and breed mosqui-toes. The latter have had a devastating impact on our native birds since they transmit bird pox and bird malaria.

LAUAʻE, MAILE-SCENTED FERN
Phymatosorus grossus
Polypodiaceae (pākahakaha, lauaʻe haole)

These are medium sized ferns up to about 2 ½ feet tall. The frond is com-posed of a single large glossy green blade, but one that is very deeply cut into long strap-like lobes that come to rounded or pointed tips. Oval rust-colored spore-bearing sori form irregular lines along each side of the midribs of the main axis and the lobes on the under surface of the frond, and often appear on the upper surface as raised round or donut-shaped rings. This fern is na-tive to New Guinea, Australia, and the south Pacific Islands, and probably to southern Asia and tropical Africa. It was first collected in Hawaiʻi in 1919. It is a common garden fern and ground cover, and has run wild in dry mesic to wet low-elevation forest on all the major Islands, often covering exten-sive areas, especially under groves of Christmas berry trees.

Some populations of this fern have a scent like that of the native maile. Traditional lore has it that lauaʻe was layered between folds of kapa to impart a fragrance to the cloth, and was used in lei. Sap was extracted from it and mixed with coconut oil

Look for: A medium sized glossy green fern with deeply cut lobes, and rust-colored oval spore clusters underneath the frond, often marked by round raised rings on the upper surface.

to scent the dyes used to stain kapa. Since this fern is a relatively recent introduction, however, it is believed that these uses actually involved a different fern, the native *Microsorum spectrum,* or peʻahi, which sometimes resembles immature lauaʻe and may have a similar scent. We think that this native was originally called "lauaʻe", but the name became applied to the more common introduced fern after it was established in the Islands.

LAUAʻE HAOLE, RABBIT'S FOOT FERN
Phlebodium aureum
Polypodiaceae (lauaʻe, pākahakaha)

This medium sized fern with 2 to 3 foot-long fronds grows on the ground or on tree trunks. The simple frond consists of a single large blade that is cut into deep, strap-shaped narrow pointed lobes and closely resembles the lauaʻe fern. However instead of being a glossy green like lauaʻe, it is a powdery bluish-green in color. The frond arises from a thick creeping stem or rhizome that is covered with golden brown hair, and looks like a rabbit's foot. The spore-bearing sori appear as oval brown bodies in an irregular row lying on either side of the midribs of the main axis and the lobes of the fern. This fern is native to tropical regions of the Americas, ranging from Florida through the Caribbean to Central and South America. It was first collected in Hawaiʻi in 1910 and has since become common. It is found in shaded mesic and wet forest up to about 2000 feet in elevation on all the major Islands.

"Haole" means foreign, implying that lauaʻe was here first and was considered a native fern, which it is not, and that this fern came to Hawaiʻi later. Thus it is ironic that this fern apparently arrived in the Islands a few years before lauaʻe.

Look for: A medium sized fern with deeply cut strap-like lobes of a bluish green color rising from a thick creeping stem covered with golden brown hair that could be imagined to be a rabbit's foot.

PALAʻĀ, LACE FERN
Sphenomeris chinensis
Lindsaeaceae (no familiar relatives)

This is a very common, easily recognized fern. It is about 2 or 3 feet tall, with light brown smooth shiny stems supporting a lacy, roughly triangular frond. The frond is subdivided 3 or 4 times with very small, elongated wedge-shaped leaflets at the end of the final stem. The leaflets are broadest at the tip, and if fertile, bear spores at the center of each tip. This species is indigenous to Hawai'i, being found also in Madagascar, southern and eastern Asia, Polynesia, and other Pacific islands. In Hawai'i it is one of the most common ferns in mesic and wet forest on all of the major Islands.

Pala'ā was often interwoven with maile to make an attractive, fragrant lei. It was sacred to Laka, the goddess of hula, and was used to decorate the altar in the hālau hula. A brownish-red dye for staining kapa was obtained from the plant. Legend tells that Hi'iaka, sister of the volcano goddess Pele, wore a skirt of pala'ā and used this to trip and entangle the mo'o (monster lizards) in her battle against them. Preparations of the fern were also used to treat ailments peculiar to females.

JOHN HOOVER

Look for: A common fern about 2 or 3 feet tall with very lacy fronds in mesic and wet forest.

PALAPALAI
Microlepia strigosa
Dennstaedtiaceae (bracken fern)

This is a medium sized indigenous fern with a twice divided frond that has a triangular shape overall, tapering to a sharp point. The soft, often hairy, bright green arching fronds are about 1 to 3 feet tall and up to a foot wide. The side stems of this pinnately compound plant bear numerous leaflets which typically have the shape of a meat cleaver—one side is relatively straight and faces the main stalk of the frond. This side is attached to the side branch by a short stem, the cleaver handle. The other side, the blade of the cleaver, faces the tip of the branchlet and bears numerous small lobes. Leaflets can be quite variable in form, with elaborate lobes however,

Look for: A bright green fern with a divided triangular frond with fairly large, cleaver-shaped, lobed leaflets in shaded moist lowland forest.

162

and some, especially near the base of the side branch, resemble small, wind-blown fir trees in shape. These may be compound themselves. The coarsely lacy side branches have the overall form of a tapering dagger blade. When fertile, the spore-bearing sori are cup-shaped bodies at the end of the lobe of a leaflet. This fern is found in lowland habitats that are moist and shady in mesic to moderately wet forest on all the main Islands. It is also native to the Himalayas, Sri Lanka, south east Asia to Japan, and Polynesia.

Palapalai was sacred to the hula goddess, Laka, and was one of the plants used to decorate her altar in the hālau hula. Hula dancers value this fern for its woody fragrance and its pliability which allows them to weave it into lei to adorn their heads, wrists, and ankles.

SWORD FERN, BOSTON FERN, KUPUKUPU
Nephrolepis spp.
Nephrolepidaceae (no familiar relatives)

There are 5 species in this genus in Hawai'i, 2 being indigenous natives and 3 naturalized aliens. They are common trailside ferns in upper dry, mesic, and wet forest. The overall shape of the fern is long and strap-like, of uniform width, tapering toward the tip like a sword, although the blade is not solid but composed of 2 rows of pinnately compound oblong leaflets, each of which

tapers to a sharp or rounded point. Tiny teeth may line the edges of each leaflet. In one of the introduced species, the leaflets may fork at the tip forming a structure that looks like the tail of a fish, giving it the name of "fishtail fern". The fronds are erect or sometimes drooping and typically about 2 feet long and 2 to 5 inches wide, although under favorable conditions they may reach several times this size. When fertile, small round sori, the spore-bearing organs, occur in a row on each side of the midrib on each leaflet. These ferns are found in mesic and wet forest on all the Islands.

Kupukupu is an early colonizer of recent lava flows on the Big Island and can be seen growing in cracks and tree holes in this environment. Introduced species can be aggressive and may form dense masses that crowd out native vegetation.

Look for: A 2 or 3 foot-tall sword-shaped fern in which each pinnately compound frond consists of a row of leaflets on either side of a stalk, and small round spore-bearing bodies lie in rows parallel to the midrib of the leaflet.

163

ULUHE, FALSE STAGHORN FERN
Dicranopteris linearis
Gleicheniaceae (uluhe lau nui)

This is a large fern. I doubt if anyone knows just how large it can grow, since the task of tracing a single plant through the matted tangle of vegetation that it forms would be virtually impossible! This fern sprawls over the ground and other vegetation, forming dense mats that quickly cover landslide scars and other disturbed

JOHN HOOVER

Look for: A sprawling, viny, constantly forking fern that forms dense mats in upper mesic and wet forest.

areas, often occupying entire hillsides, precipitous slopes, and vast areas in the upper mesic and wet forest. If you hike into mesic or higher forests in Hawaiʻi, you will not be able to avoid this fern! In dry areas, the fern mats may be 1 or 2 feet thick, but in moister country they may be 12 feet thick or more and become a very serious obstacle to travel. Uluhe stems are round, dark brown, brittle, and repeatedly fork. Each bare stem ends in a pair of smaller stalks which bear 2 rows of oblong leaflets forming a blade-like structure that tapers to a slender point. Small round spore-bearing sori may be seen forming short and broken rows parallel to the midrib of leaflets on older portions of the plant. Uluhe is found on all the higher Islands. The fern is indigenous, being found also in tropical Africa, Asia, Australia, New Zealand, Japan, America, Polynesia, Indonesia, and the Caribbean islands.

Uluhe provides a valuable service in covering bare ground. The dense mats and especially the thick tough layer of stems and rhizomes and dead leaves that cover the ground beneath the mats cushion the impact of raindrops and help prevent erosion. The fern has been measured to invade fresh scars of bare earth from the edge at a rate of 3 to 7 feet per year. It is a constant struggle to keep trails open where they pass through uluhe-covered areas! The Hawaiians drank an infusion brewed from uluhe fronds as a laxative. One of our native damselflies, a group that normally spends its larval days in fresh water, lays its eggs in uluhe mats and the larvae develop in the damp leaf litter beneath the fern mass. A sea bird, the Newell's Shearwater, digs burrows under uluhe mats on steep wet mountain slopes and lays eggs in these. The adult birds forage out to sea during the day and return to feed the chicks and incubating mates in the night. I remember being awakened late one evening in the summer of 1977 by the strange calls of these birds returning to their nests, when my friends, Fred, Charlie, and Alyce Dodge, and I were making a madcap exploration across the north slope of the North Kohala Mountains near the rim of Waimanu Valley. If you hike in wet forest, you will occasionally see a relative of uluhe, that the Hawaiians called "uluhe lau nui" or "large-leaf uluhe". This plant bears an obvious resemblance to uluhe but does have markedly larger fronds and much more limited growth, rarely covering large areas or forming thick mats.

MISCELLANEOUS

LICHENS
Teloschistes flavicans
Teloschistaceae
Usnea australis
Parmeliaceae

There are two conspicuous lichens that you are likely to notice on our trails. They are very similar in form, but differ in color. Both are tufts of tangled filaments, an inch or two in diameter, and are often seen on exposed branches or dead twigs. They may occur on rocks also. These fascinating symbiotic organisms, a cooperative form of life in which a photosynthetic cyanobacterium or an alga, and a fungus, live together to form a single, highly successful life form, are not parasitic, but perch on the twig in order to reach the light that the light-harvesting partner needs for photosynthesis. They will be found most commonly in mesic forest. The orange colored one is *Teloschistes flavicans*, while the gray is *Usnea australis*.

Look for: An open, filamentous tuft, like a fragment of well-used scouring pad, colored gray or orange, in an open area on the twig of a tree.

Thomas H. Rau

LIVERWORTS

Not many people are likely to be familiar with liverworts, although they are quite common along our trails and are of considerable interest. There are two main groups of liverworts, the ribbon-like thalloid liverworts and the much more common leafy liverworts. The latter look much like mosses, but can be distinguished from them by several features. Most mosses have their leaves arranged in a spiral around the stem while all leafy liverworts have two major rows of leaves that lie in a plane on opposite sides of the stem with adjacent leaves overlapping like the shingles on a roof. Mosses also often display a spore-bearing fruiting capsule, a small brown oval body on top of a long slender stalk. There is almost never an obvious fruiting body on a stalk on a leafy liverwort.

Liverworts and mosses differ from ferns and higher plants in lacking a vascular system that would allow them to transport water from the soil to distant leaves. All parts of the plant must be in close contact with moisture, therefore. This means that it is necessary for them to live in very wet environments or to survive in a dormant state while being dried out for extended periods of time.

The ribbon-like thalloid liverworts that you are likely to see are confined to quite wet sites where they cling to damp rocks or soil banks in shady dark gulches, or in the spray zones near a waterfall. An example of this kind of liverwort is *Dumortiera* which is a gray-green ribbon about ¼ inch wide that branches into 2 equal tongues, each with a broad, rounded tip that usually has a dimple in the center as if it were getting ready to branch again. Often you

Look for: A gray-green ribbon that forks into 2 equal branches and often bears a miniature umbrella near the tip.

will see a small umbrella-like structure rising from the base of this dimple. This is the reproductive, spore-bearing, structure of this liverwort.

If you carefully examine the masses of "moss" that grow on tree limbs and trunks in the wet forest you will notice that many of these are composed of slender, branching filaments about ⅛ inch in width that bear rows of tiny leaves that overlap like shingles. These are leafy liverworts. They may be green but often have a distinctly reddish hue or even a darker brown or blackish color. They grow in dense masses on the trees or ground intermingled with mosses, ferns, and other small plants.

JOHN HOOVER

My favorite among the liverworts however, is a hardy group of tiny plants in the genus *Frullania*. These rugged individuals can be found in mesic forest and even down to the verge of the dry forest where they grow on the bark of a variety of trees such as Java plum, 'ōhi'a, and koa. In favorable locations they may grow in dense, black masses, but are seen to best advantage where they are spread out and the delicate tracery of their tiny dark filaments

Look for: Slender branching filaments bearing pairs of overlapping leaves in a shingle pattern, growing on trees in the wet forest.

is exposed against the gray bark of the host tree. Here the pioneering tendrils, often ⅛ of an inch across or less, create beautiful tree-like or fern-like patterns that are well worth noticing and admiring! Look closely! Think small!

Another group of leafy liverworts that are worth mention are those that grow on the leaves of other plants, such as mountain apples or rose apples. If you examine the oldest leaves of these trees, you will occasionally notice a few filaments with overlapping shingle leaves pressed to the surface of the tree leaf. These little liverworts are not parasites, but are simply using the leaf of the higher plant as a surface to grow on that raises them above the forest floor and allows them to reach more light. It is necessary, of course, for them to carry out their entire life cycle during the short life span of one leaf. Liverworts are thought to contain a wide variety of unusual chemicals, although we know little about these as yet, and they are being investigated. We know that dried specimens are

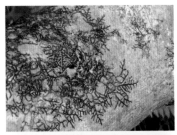

Look for: Wonderfully detailed fern-like patterns in black against the gray bark of a tree in mesic forest.

seldom attacked by insects in herbaria, and suspect that they contain insect repellant or insecticidal compounds. Medieval Europeans thought that liverworts contained a treatment for liver ailments. According to the Doctrine of Signatures, they believed that God had placed a treatment on earth for every human disease. The plant containing the treatment for a malady would resemble the organ affected by the disease! Thus they thought that the thalloid liverworts looked somewhat like a liver and would contain medicines effective against hepatitis and other liver conditions. Hence the name "liverwort" in which "wort" is simply an old word for "plant".

MOSSES

Everyone is familiar with mosses, although not all of them *are* mosses—see the entry on LIVERWORTS. In the wet forest the trees bear a thick coat of mosses, liverworts, ferns and other small plants so that the diameter of a limb often seems to be several times its actual size. Even in the upper mesic forest some trees may have thick clumps of moss where the tree forks or rain collects, and these provide a foot-hold for ferns and the seedlings of ʻōhiʻa and other trees, which often get their start in life perched on another tree. Not all mosses

Look for: Thick, compact layers of a pale green moss along trails in shady mesic forest.

Look for: A compact moss clinging to tree limbs, made up of many uniformly sized fuzzy balls with octopus-like tendrils extending from the main colony.

BRADLEY F. WATERS

JOHN HOOVER

Look for: A large moss, often on fallen logs, that looks like a dense mass of tiny fox tails.

require a great deal of moisture. One that you are likely to notice along trails in mesic forest is *Leucobryum* which forms a dense, pale green to almost white turf with a surface like coarse terrycloth, that covers extensive areas of soil beside the trail. It seems to favor trailsides and on seldom used paths, will even form the surface of the trail itself.

Another of the many mosses that you might encounter is *Pyrrhobryum spiniforme*, which often forms a dense growth on fallen logs in moist mesic and wet forest. This is a fairly large moss with stems an inch or so in length surrounded by hair-like leaves, so that the plant resembles a miniature fox tail.

One final moss that I will mention is *Macromitrium piliferum*, a common moss seen on tree trunks, limbs, and rocks in fairly moist mesic and wet forest. This moss is rather small and compact and looks like a collection of small fuzzy balls (the pills of piliferum?). The edges of the clump often send out octopus-like tentacles of new growth.

PEPEIAO
Auricularia cornea
Basidiomycetes (mushrooms, bracket fungi, stinkhorns)

Both the Hawaiian and the genus names mean "ear", and this is appropriate for this ear-like irregularly cup-shaped, thin brownish-gray fungus which is found attached by an edge or the base of a cup to

Look for: Flexible rubbery ear-shaped gray-brown fungi growing on decaying logs in moist gulches.

a decaying log. The surface of the fungus may be smooth or wrinkled. The flexible rubbery body is about ⅛ of an inch thick. It is often found growing in clusters on a fallen kukui log in damp gulches in mesic forest.

This well known jelly fungus is collected by local people, dried or used fresh, cut into thin strips, and added to a variety of dishes, such as chicken hekka. It has little flavor itself but adds an interesting crunchy texture to the dish and soaks up the flavors of the other foods.

POLYCHROME WOOD FAN
Microporus flabelliformis
Basidiomycetes (pepeiao, mushrooms, puffballs, shelf fungi)

This fungus grows as a semicircular plate, often with irregular or lobed edges and radial ridges, on decaying wood. It is tough and leathery, up to 3 inches or a bit more in diameter and about ⅛ of an inch thick. It often occurs in clusters. The fruiting body (the part you see) is often beautifully patterned with concentric arcs in brown, reddish, yellowish, gray-green, orange, and black with a rim of white around the outer edge. It is common in mesic and moist forest on most of the major Islands.

Look for: A colorfully patterned leathery fungus, shaped like an open fan, growing on decaying wood.

Crafts workers in Hawai'i often collect and dry this fungus and incorporate it into earrings, necklaces, and other bits of art work.

SCARLET WOOD FAN
Pycnoporus sanguineus
Basidiomycetes (stinkhorns, mushrooms, pepeiao)

This fungus forms semicircular plates up to 3 inches in diameter and ⅛ to ¼ inches thick on decaying wood. It is leathery and pliable and the upper surface feels slightly velvety. In color it ranges from bright red to orange-red with paler bands parallel to the curving rim, but may fade somewhat with age. Frequently it occurs in small clusters. It is common in mesic and moist forest on all the major Islands. Weavers extract a yellow dye from this fungus.

Look for: A red or red-orange fungus shaped like an open fan, found on decaying wood.

TRENTEPOHLIA
Trentepohlia aurea
Trentepohliophyceae

This tiny plant is classified as a filamentous green alga, even though you will see it as being Chinese red in color. It is conspicuous in our mesic forest where it often gives a reddish tinge to the bark of koa and other trees. It also coats rocks in favorable locations, providing a rusty hue to one of their faces. For awhile, it grew abundantly near the top of the Kuliʻouʻou Trail on Oʻahu where it covered the limbs of practically all the trees and shrubs with a fuzzy reddish coat. As many of the trees seemed to be dying, hikers suspected that it was some kind of disease that was killing them. Actually, the dairy farm in Waimānalo at the foot of the pali here, whose effluvia could easily be detected when the trade winds were right (most of the time) supplied a detectable amount of ammonia to the air, and this probably fertilized the alga and stimulated its growth. It may also have been strong enough to poison some of the trees and contribute to their die-back. The alga itself is harmless to its hosts, which simply supply a surface on which it can grow and reach the light. The dairy is gone now, as is the trentepohlia in this area, but a less vigorous growth of this alga can be seen on many of our dry to mesic forest trees and shrubs.

Look for: A reddish tinge to the bark of a tree or on a rock. Sometimes this is thick enough to form a fuzzy layer.

HERBS

ARTHROSTEMA
Arthrostema ciliatum
Melastomaceae (clidemia, glory bush, miconia)

KEN SUZUKI

This is a large lanky sprawling herb with weak succulent stems that are square in cross section. The stems fork repeatedly in pairs, but there are often long sections of stem that have no major branches. The canes may be 3 to 12 feet long. The oval pointed leaves are 1 to 3 inches long and have 5 to 7 major veins radiating from the base toward the tip. Attractive pink flow-

Look for: A sprawling lanky herb with weak succulent square stems, leaves with 5-7 major veins radiating from the stem, and pink flowers.

ers a bit more than an inch in diameter are composed of 4 well separated petals. The fruit is an oblong body about ⅝ of an inch long with a square crown at the tip. The plant is native to tropical America and was first collected on Oʻahu in 1939. Arthrostema is occasionally planted in gardens and may have been introduced for this purpose. It has recently begun to spread and is now a nuisance on a number of our trails in wet forest and moister parts of the mesic forest on Oʻahu. It also occurs on the Big Island and perhaps others as well.

ASIATIC PENNYWORT
Centella asiatica
Apiaceae (carrot, parsley, dill, coriander)

JOHN HOOVER

This plant is a small creeping herb with rounded, slightly scalloped, kidney-shaped leaves ¾ inch to 2 inches in diameter. The leaves spring in clusters from a common point and runners root periodically and send up additional clusters. The plant produces small white

173

inconspicuous flowers which lie so close to the roots that they are partially underground and not often seen. The plant is native to Asia where the leaves are eaten and used medicinally. It was present in Hawai'i as early as 1871, perhaps reaching the Islands as a contaminant of seeds of other Asian plants. It is now common in moist areas, including low damp areas in drier forest, on all the larger Islands.

Look for: A low, creeping herb with rounded kidney-shaped leaves.

DAISY FLEABANE
Erigeron karvinskianus
Asteraceae (silversword, sunflower, lettuce)

These plants are sprawling herbs that often form dense clumps. The variable alternate leaves range in form from elongated smooth-edged ovals to broad wedges with serrated edges and 3 pointed lobes. They are about an inch long and ½ inch wide. Single flowers are borne on slender stems and have 60 to 70 thin white aster-like petals that turn pinkish with age. The flower is about ¾ of an inch across and has a yellow center. The plant is native to Central and South America, from Mexico to Chile, and the Caribbean Islands. It was being cultivated as an ornamental in Hawai'i by 1911, and now grows wild in moist areas on all the major Islands. The name "fleabane" implies that it may have insect repelling properties.

KEN SUZUKI

Look for: A low sprawling herb with white aster-like flowers that turn pinkish with age.

GINGERS
Hedychium spp.
Zingiberaceae (ornamental gingers, spice ginger, 'awapuhi)

Three recently introduced gingers have become fairly common along trails in wet disturbed areas. A fourth, 'awapuhi, was introduced by Polynesian settlers and is considered separately.

Yellow ginger, *Hedychium flavescens*, sends up straight, unbranched shoots 3 to 6 feet tall with leaves alternating on opposite sides of the stalk. The leaves are long pointed blades about 12 to 18 inches long and 2 to 4 inches wide. Flowers, borne on the top of the stalk, are 2 or 3 inches wide, yellow, and fragrant. There is a large upper petal with 2 more slender petals below it and 3 thin filaments. The plant is native to northeast India and the Himalayas and was probably introduced to Hawai'i by Chinese immigrants as an ornamental in the late 1800s. It was first collected here in the wild in 1913.

Look for: A tall, large-leaved herb with fragrant yellow blossoms having 1 prominent upper petal, two narrower wings, and 3 slender filaments.

White ginger, *H. coronarium*, is identical to yellow ginger except that the flower is white. This plant is probably native to southern China and the Himalayas and was also introduced as an ornamental, probably by the Chinese, prior to 1888.

Look for: A plant very similar to the one above, but with white flowers.

Kāhili ginger, *H. gardnerianum*, has a stalk much like the 2 above, but many yellow flowers with bright red-orange stamens are in blossom at one time on this plant and they are arranged in vertical rows, 60 degrees apart, in a cylinder around the central stem. Looking down on this cluster from above, you can see that the rows form a perfect hexagonal array, each row being a radius from the stalk to one corner of a regular hexagon. The plant produces bright red seeds which are attractive to birds and are thus widely spread into native rainforest. This ginger is also native to the Himalayan region and was first collected around 1940 in Hawai'i Volcanoes National Park, where it was planted around the Park housing area as an ornamental and has now become a major pest.

THOMAS H. RAU

Look for: A plant like those above but with a cylinder of yellow flowers in hexagonal array along the terminal stalk.

All of these plants can be a menace to our native ecosystems. The white and yellow gingers rarely produce seeds and so tend to spread vegetatively along roadsides and streams from sites in which they have been planted. None the less, they have become a serious pest in some places, especially around Kōkeʻe on Kauaʻi. Kāhili ginger is a still more serious problem. Like the others, it produces dense mats of underground stems (rhizomes) which crowd out all other plants. It is highly tolerant of shade, and since its numerous seeds are spread by birds, it is capable of rapid dispersal, and is an aggressive invader of native rainforest, where it may virtually eliminate all other under-story plants and prevent the growth of tree seedlings.

HAIRY CAT'S EAR, GOSMORE
Hypochoeris radicata
Asteraceae (sagebrush, thistle, dandelion)

This hairy weed from the Mediterranean area grows as a low rosette with fibrous roots. The leaves are about 1 to 10 inches long by ½ to 2 inch wide with large, lobe-like teeth along the sides. Branched or unbranched leafless, slender stalks bear yellow flower heads about 1 inch in diameter that look much like the flowers of the common dandelion of the mainland United States. Like the dandelion it produces tiny seeds, each with its own parachute, that allow the wind to carry it for great distances. The plant was introduced into Hawaiʻi sometime before 1909 and has

Look for: A hairy, rosette-forming weed with dandelion-like leaves and flowers.

adapted to a wide variety of conditions, being found in wet to relatively dry sites at elevations from sea level to near the summits of our highest mountains on all the major Islands except Niʻihau and Oʻahu.

A close relative, the smooth cat's ear, *H. glabra*, from Europe, grows in dry sites on all the Islands. This plant has a tap root and much smaller flowers, about ¼ inch in diameter, but where the two occur together they may hybridize and produce a swarm of plants with intermediate forms.

KNOTWEED
Polygonum capitatum
Polygonaceae (buckwheat, rhubarb, sea grape, Mexican creeper)

Seven weeds in this genus have been introduced into Hawaiʻi. The species above is the one you are most likely to see along our trails. It is a low-growing, mat-forming herb with broadly spindle-shaped leaves an inch or two long that bear a reddish-purple V-shaped band. The whole plant is often a reddish color when exposed to full sunlight. The most conspicuous feature of this plant is the flower cluster which consists of a pinkish-colored ball covered with tiny but prominent knobs. The plant is native to the Himalayas and western China and has been introduced into Hawai'i as an ornamental, primarily as a ground cover. It was first reported on the Big Island in 1960, and is now found on most of the major Islands along roadsides, in wet forest, and on open lava fields.

Look for: A mat-forming herb with a purplish V-shaped band across each leaf and small round knobby pinkish flower clusters on slender stalks.

MACHAERINA, 'UKI

Machaerina spp.
Cyperaceae (papyrus, nut grass, great bulrush)

Look for: A large, dense, clump of glossy green strap-like leaves growing on the ground in the open in wet forest.

Look for: An attractive bluish colored sedge with leaves that emerge in alternation in a plane from the base of the plant, and, in season, bearing an open tuft of small brown flowers.

Two species of machaerina are indigenous to Hawai'i, being found also on islands to the south. The Hawaiians called both of these species 'uki, but since they also used that name for several sedges and for the native lily, *Dianella*, it seems better to use the scientific name to avoid confusion. *M. angustifolia* is an unmistakable feature of the wet forest. It grows in large clumps with many long smooth glossy strap-like leaves

177

about an inch wide, with no midribs, rising from the base of the plant. These leaves may be 3 or more feet long. The flowers consist of small brown tufts arranged along thin stems that branch from a tall stalk to form an open elongated oval cluster. This plant is found on ridge tops in wet forest on all the higher Islands and is characteristic of the wet forest.

Pigs may tear clumps apart to feed on the succulent leaf bases. The flowering stalks of this plant were used as "hair" to decorate the gourd masks worn by priests during the makahiki festival of Lono, when tribute was collected from each ahupua'a or district.

Machaerina mariscoides is a smaller sedge with leaves that taper to a slender point and have an attractive bluish color. The leaves form a nice pattern as they emerge in overlapping alternation in a plane at the base of the plant. The flower stalk forms a loose brown oval, much like that of the plant above, but is borne on a zigzag stem. This plant is found in upper mesic and wet forest on all the higher Islands.

ORIENTAL HAWKSBEARD
Youngia japonica
Asteraceae (zinnia, cocklebur, aster, na'ena'e)

This plant is a low growing herb with most leaves forming a rosette at the base. Leaves range from 1 to 8 inches long by ½ to 2 inches wide with pointed lobes along the sides, a pointed tip, and an overall paddle-shape, being broadest toward the tip. Flowers are borne on slender, branched stalks and are small, about ¼ inch across, yellow, and aster-like in form. The plant is native to southeast Asia but is now widely distributed as a weed throughout the tropics. It was first collected in Hawai'i in 1864 or 5, and is now common in shady damp sites on all the major Islands.

Look for: A rosette-forming herb with small yellow aster-like flowers on slender branching stalks.

PA'INIU
Astelia spp.
Liliaceae (lilies, amaryllis)

Three species of this endemic lily-like plant are found in Hawai'i. Occasionally they will be found growing on the ground in wet forests, but because they are eaten by pigs, you are more likely to see them perched on logs and tree limbs out of reach of these voracious pests. The long slender pointed leaves rise in alternation from the base of the plant and are covered with silvery hairs, especially underneath. The leaves usually have 3 prominent veins. Small inconspicuous 6-petalled lily-like flowers in a variety of colors are produced in clusters on a stiff silvery hairy stalk. These are followed on female plants by clusters of attractive orange berries which birds favor, helping to spread the seeds.

JOHN HOOVER

Look for: Tufts of slender silvery sedge- or lily-like leaves growing on low limbs of trees in the wet forest and sometimes bearing clusters of orange berries.

Early Hawaiians occasionally used the lovely silvery leaves of this plant in lei.

TARO, KALO
Colocasia esculenta
Araceae (monstera, philodendron, calla lily, caladium, anthurium)

Taro was introduced to the Islands by the first Polynesian settlers and was the preferred staple of the Hawaiian peoples. It can still be found occasionally growing wild along mountain streams, although in this habitat it is usually much smaller than the cultivated varieties. Taro has large heart-shaped leaves perched on a cylindrical stalk that rises from the base of the plant. The stalk is not attached to the margin of the leaf, in the usual fashion, but to its back, a little below the end of the deep cleft at the top of the leaf. From the front, the point of attachment can be seen as a small dimple, or piko in Hawaiian, which was the name then given to the umbilicus, or navel, in humans.

The importance of taro to Hawaiian culture can be seen in the Kumulipo, the creation chant that explains the origin of humankind. There are several versions of the story, but they all agree that Wākea, the sky god or principle, who was the male and was sometimes identified with Kāne, mated with Papa, the earth, the female. From this union (or from his second mating with the daughter that resulted from this one) a child (or perhaps a root) was born. This child was deformed or stillborn and was buried behind the house. A taro plant grew from the grave. This was named Hāloanaka or Long Stalk Trembling Leaf. A second mating gave rise to a viable son, who was named Hāloa, or Long Stalk, and was the ancestor of all humans. Thus according to Hawaiian mythology, the taro plant was the elder brother of humankind, and as such was more sacred and entitled to more respect than humanity itself. King Kalākaua traced his ancestry back to the second Hāloa, and used the taro leaf symbol on his crown. Because of its high sacred status, cultivation and preparation of taro was kapu, forbidden, to women, and only men were allowed to grow, harvest, prepare, and cook the staple. Once poi was prepared from the corm however, women were allowed to eat it—it was not one of the many foods forbidden to them.

By far the most important use the Hawaiians had for taro was as their staple food, usually in the form of poi. Most Westerners seem to find poi rather unpalatable—even my locally-born family prefers to eat it with sugar or honey. I must confess that I rather like it myself—it is bland, but no more so than rice or potatoes. Taro found many other applications as well, being used as a glue to help bind strips of kapa together, in fish bait, and as a component in a variety of medicines. All parts of the plant contain sharp crystals of calcium oxalate which are extremely irritating to the membranes of the mouth, so that thorough cooking is necessary to dissolve these and make taro edible. Medicines were usually prepared with scrapings of the raw corm, however, perhaps with the idea that medicine is most effective when it is most nasty! Probably varieties that were lowest in the crystals were used for this purpose. In any case, different parts of the plant were included in concoctions to treat asthma, thrush, heart burn, insomnia, and to ease childbirth.

Look for: A low growing stream side plant with light green, heart-shaped leaves attached to a stalk that joins the leaf at the back and inside, rather than at, the margin.

SHRUBS

ʻĀKŌLEA
Boehmeria grandis
Urticaceae (māmaki, nettles, pilea)

This endemic shrub is about 3 to 6 feet tall and tends to sprawl. It has large opposite leaves that are broadly spindle-shaped and about 5 to 10 inches long by 4 to 6 inches wide. The leaves are toothed along the edges and have 3 prominent red veins extending from the base toward the end of the leaf. Māmaki leaves may be very similar, but they are alternate, and the fruits of these two plants are quite distinctive. The flowers and fruit are borne on long, slender, branching strands up to 6 or 8 inches long that grow from the base of the leaf stem. Flowers appear as tiny round clusters along these filaments. The shrub is found in moist mesic valleys and wet forest areas.

JOHN HOOVER

Look for: A sprawling shrub with large, toothed, spindle-shaped leaves with 3 prominent reddish main veins and long branching filaments bearing small round flower clusters.

ALANI
Melicope (Pelea) spp.
Rutaceae (citrus fruits, rue, mokihana, mock orange)

There are 47 endemic species in this genus! They are trees or shrubs, generally of the upper mesic or wet forest, with opposite or whorled leaves. The leaves tend to be about 3 inches long, oblong, with blunt or indented tips, although in a few species they are pointed. The midrib is usually quite prominent, and often dark in color, but the lateral veins are very fine. The upper leaf surface is glossy. Leaves tend to be stiff and brittle with the edges curling under, and when crushed a few species give out a strong anise scent. Others may have little odor, or smell vaguely of citrus or turpentine. Leaves often have prominent knobs on the surface—galls formed by a kind of mite, a psyllid (insect), or a fungus, any of which may attack this plant. Flowers occur on short stems, singly or in small clusters next to the twig, and have 4 small whitish petals. The green fruit, about ⅜ inch across, is divided into 4 distinct lobes,

181

each forming the corner of a square, and are an unmistakable clue to the identity of this plant. Most species are restricted to very limited areas on the Islands, and many are probably extinct, but one kind of alani or another will be frequently encountered, particularly in wet forest, on all the higher Islands.

Look for: A shrub or small tree of wet areas with stiff brittle oblong leaves that curl under along the edges and 4-lobed fruit, each lobe at the corner of a square.

The Hawaiians seem to have made little use of these plants, with the exception of the moki-hana, *M. anisata*, which is found only on Kaua'i. All parts of this alani have a strong anise scent, and the berries and leaves were placed among stored kapa to perfume it. The berries were also strung into fragrant lei, which would retain their scent for a long time. Note however, that some people are allergic to compounds in the berry and may develop a rash after wearing such a lei. I have occasionally found alani on O'ahu with anise-scented leaves, but this is unusual. A related genus, *Platydesma*, with 4 endemic species, is similar, but has significantly larger flowers and fruit.

CLERODENDRUM, PĪKAKE HOHONO

Clerodendrum chinense (philippinum, according to the *Manual)*
Verbenaceae (lantana, fiddlewood, Jamaica vervain, verbena, vitex)

These introduced ornamentals are shrubs 6 feet or more tall with quadrangular branches. The leaves are broad, roughly triangular with rounded corners, irregular edges, and prominent veins. They are opposite, with each pair emerging at right angles to the previous pair. The leaves are reported to have a foul smell when bruised. The fragrant white double flowers are tightly clustered at the ends of the

Look for: A shrub of wet, shady areas with opposite, rounded-triangular leaves that have a foul smell when crushed and terminal clusters of fragrant white double flowers emerging from a trumpet-shaped maroon base with 4 to 6 slender points.

branches, and the buds are a deep pink at the tips. Each flower emerges from a maroon-colored trumpet-shaped receptacle with 4 to 6 slender points, giving the whole cluster a maroon and white coloration. We are not sure of the original home of this plant—probably China or south Asia—but it is now widespread throughout Asia, the Pacific, Africa and the Americas, being widely cultivated as an ornamental. The double-flowered variety does not set seed, but spreads readily by root suckers. It was first collected in Hawai'i in 1864 or 5 and is now naturalized in open wet shady areas at low elevations on all the larger Islands.

GLORY BUSH, LASIANDRA, TIBOUCHINA

Tibouchina urvilleana
Melastomaceae (clidemia, miconia, false meadow beauty)

Glory bush is a shrub or small tree up to 12 feet tall. The young stems are square in cross section and hairy. The leaves are pointed ovals about 2 to 5 inches long by 1 to 2 inches wide, and have 5 to 7 prominent veins running from base to tip, as is typical for this family. The 5-petaled flowers are 3 inches in diameter and a deep, velvety purple in color. Unopened buds are enclosed in bright pinkish red sheaths, and the attractive

JOHN HOOVER

Look for: A wet forest shrub with bright pink buds and deep velvety purple flowers 2 to 3 inches in diameter.

leaves have a silvery sheen due to numerous green hairs. Older leaves turn a bright scarlet and compete with the flowers for attention. In all, a very ornamental plant. Unfortunately, it can form dense thickets that crowd out natives and all other plants. Glory bush is native to southern Brazil and was probably introduced into Hawai'i about 1910, although first reports of its presence in the wild were in 1917. It is now naturalized and spreading in cool wet forests on all the larger Islands, and particularly around Kōke'e on Kaua'i and along the highway between Hilo and Volcanoes National Park on the Big Island.

HAʻIWALE, CYRTANDRA
Cyrtandra spp.
Gesneriaceae (African violets, gloxinia)

The *Manual* lists 53 endemic species in this genus, the most for any genus of plants in Hawaiʻi. The plants are so variable, however, that it is often difficult to decide exactly what a species is in this group and there is a great deal of confusion about their classification, with many plants appearing to combine features of supposedly different species. In wet, sheltered gulches haʻiwale may have large, nearly round, bluntly pointed, soft hairy velvety leaves, while those found at the top of a wind-swept pali (cliff) may have stiff narrow corrugated sharply pointed leaves. However, once you have seen the flowers and fruit of haʻiwale, one or the other of which are usually present, you will always be able to recognize these plants. The haʻiwale of the gulches is an open, rather gangly slender shrub with soft wood. The leaves are opposite or whorled. Flowers are trumpet-shaped with 5, often unequal, lobes, with the lower ones usually being the larger. The blossoms are white, about ½ inch across, and usually attached to one of

Look for: A wet habitat shrub with white, trumpet-shaped flowers with 5, often unequal, lobes, and white oval berries.

the main stems of the plant, sometimes near the ground or even on exposed roots. The fruit is a white oval berry about ¾ of an inch long. Whether found in gulches or on ridge tops, haʻiwale is always a plant of moist habitats. It is found on all the Islands high enough to have wet gulches.

NIGHT CESTRUM
Cestrum nocturnum
Solanaceae (tomato, potato, tobacco, chili pepper)

These shrubs or small trees are sprawling, white barked, invasive weeds of upper mesic and wet forest. The alternate, lance-shaped leaves are about 4 to 6 inches long by 1 ½ wide. Clusters of small flowers on slender branching stalks emerge from the twig at

Look for: A sprawling, light-barked shrub with clusters of small tubular flowers or white berries, particularly on the trails around Tantalus on Oʻahu.

the end or at a leaf base. Each flower is a slender greenish-yellow tube about ¾ inch long, that expands toward the tip and ends in 5 small pointed petals like a tiny crown. These are followed by white berries about ⅓ inch in diameter.

Three species in this genus have been introduced to Hawaiʻi as ornamentals. They are native to the West Indies and Central America and were first cultivated here prior to 1871. The species above has established itself and become a serious pest in the Tantalus area and has also become naturalized in the Hāʻena area on Kauaʻi. Orange cestrum, *C. aurantiacum*, which has orange-yellow flowers and white berries, is thought to be naturalized on Oʻahu, Maui, and Hawaiʻi. Day cestrum, *C. diurnum*, with white flowers and black berries, seems to be established in the wild on Kauaʻi, Oʻʻahu, and Molokaʻi.

OLOMEA
Perrottetia sandwicensis
Celastraceae (khat, false olive)

This endemic species is a shrub or small tree with shiny dark green alternate leaves that can always be recognized by the reddish leaf stems and veins. The leaves are pointed ovals, about 3 to 6 inches long by 1 to 3 inches wide with toothed edges. Olomea is a fairly common understory plant of wet forests on all the major Islands.

The early Hawaiians seem to have favored the wood of this shrub for the fire plow used to light their fires. A stick of hard olomea wood would be rubbed or twirled against a block of softer wood, such as hau, until the friction generated enough heat to cause the resulting sawdust to smolder and catch fire.

Look for: A shrub or small tree of the wet forest with shiny, toothed leaves that have stems and main veins that are red in color.

185

OLONĀ
Touchardia latifolia
Urticaceae (māmaki, nettle, artillery plant)

KEN SUZUKI

Most of the plants that were valued by the Poly-nesians were carried with them from island to island as they spread throughout the Pacific. Olonā is perhaps the only endemic Hawaiian species that proved to be a significant addition to their cultural kit. This plant is a leggy, straight-stemmed, little-branched shrub up to 10 feet tall with large alternate leaves that are

Look for: A leggy, sparsely branched shrub up to 10 feet tall with large, broadly spear-head shaped leaves at the top of each stalk.

shaped like broad spear-heads. The leaves are 8 to 20 inches long by 3 to 8 inches wide with straight sides converging to sharp points. Olonā is found in wet, sheltered gulches in mesic to wet forest on all the larger Islands. It is not a common plant and can be most readily seen in one or two small gulches right above the Kalawahine Trail on Mt. Tantalus, Oʻahu.

The inner bark of olonā could be processed into the world's strongest natural fiber—about 8 times as strong as hemp. It was white, flexible, and did not kink. An olonā line did not deteriorate in sea water, and when properly spun, was a thing of beauty. The Hawaiians used it for a great variety of purposes for which thread, string, or rope might be useful. These included tying off a baby's umbilical cord, lei strings, fish nets, fish line, carrying nets, the network on which feathers were tied to make the wonderful feather cloaks, feathered gods, or war helmets of the most noble aliʻi, the net base of ti leaf rain coats, thread to stitch kapa together, cords to bind adze heads to their handles or shark's teeth to a club, and a myriad of other uses. When western whalers and merchants reached the Islands, they valued olonā cordage very highly for rigging their sailing ships. In the 1870s King Kalākaua required that a portion of the taxes due him be paid in olonā rope which he then sold at a high price to Swiss alpine clubs for climbing ropes, and olonā continued to be prized for this purpose until nylon and other synthetics were developed in the mid 20th century.

OXYSPORA
Oxyspora paniculata
Melastomaceae (clidemia, glory bush, miconia)

This plant is a sprawling shrub with branches up to 6 feet long, with opposite leaves that are broad pointed ovals 4 to 10 inches long by 2 to 4 inches wide. Five to 7 prominent veins run from the base to the tip of the leaf, as in other melastomes. The young branches are hairy. Clusters of 4-petaled pink to purple flowers hang down in pyramid-shaped clusters from the tips of the branches, in season. The plant is native to the Himalayan region and has been cultivated in gardens in Honolulu.

Look for: A plant that looks like a larger softer more sprawling clidemia. Oxyspora has larger and smoother leaves than clidemia but a similar vein pattern.

It was first collected in the wild in 1954 and is now spreading in wet areas in the mountains above the City, being found in Nuʻuanu Valley, Mānoa Valley, and the adjacent hills. The Hawaiʻi Department of Agriculture classifies it as a noxious weed which should be eradicated wherever found.

PŪʻAHANUI, KANAWAU
Broussaisia arguta
Hydrangeaceae (hydrangea)

This endemic hydrangea is a shrub up to 15 feet tall. The large leaves are opposite or in whorls of 3, coarse, with prominent indented veins, and teeth along the edges. The leaves are broad pointed ovals about 4 to 10 inches long by 1 to 4 inches wide. They leave prominent half-round leaf scars on the twig. Flowers occur in compact umbrella-shaped clusters at the top of the plant. They vary in color from cream or yellow to pink or bluish, but the petals are small and not very showy. The oval, half-inch long fruit has a reddish hue. Male and female flowers occur on different plants. Pūʻahanui is common in the wet and upper mesic forest on all the larger Islands.

JOHN HOOVER

Look for: A shrub of the moderately wet forest with large coarse toothed leaves and umbrella- shaped clusters of flowers.

TREES

KALIA
Elaeocarpus bifidus
Elaeocarpaceae (blue marble tree)

This endemic tree grows to about 30 feet in height. It has alternate leaves, usually clustered near the ends of the twigs, that are 3 to 7 inches long and about 2 inches wide and come to a sharp point. They usually bear slight serrations along each edge. The leaf stems may be yellow or even orange. The end of the twig where the leaves emerge is slightly sticky to the touch. Flowers and fruit are only seen occasionally on this tree, and a mite often damages the flowers, causing them to form a red witches broom. This

Look for: A small tree of upper mesic and wet forests on Oʻahu and Kauaʻi that has alternate serrated, pointed leaves that are usually mottled yellow and have rust spots.

tree usually looks rather unhealthy. Only the youngest leaves are likely to be solid green, while older leaves become mottled with yellow and develop rust-colored spots. Kalia is not uncommon in the upper mesic and wet forests on Oʻahu and Kauaʻi.

A member of the same genus, *E. grandis*, from Australia, has been planted as an ornamental in Hawaiʻi. It produces fruit of an unusual blue color about the size and shape of a large marble. After the flesh decays, a delicate, intricately patterned woody capsule remains that makes an attractive bead for a necklace or craft work. These trees are most easily seen on the Pauoa Flats trail on Mt. Tantalus, Oʻahu or the University campus in Mānoa.

KĀWAʻU
Ilex anomala
Aquifoliaceae (English holly, yerba mate or Paraguay tea)

This indigenous native is also found in Tahiti and the Marquesas. It is a small to medium sized tree of the upper mesic to wet forest. The leaves are dark green and slightly glossy on the upper surface, while the lower is paler and not glossy. Leaves are oval with blunt tips and about 2 to 6 inches long, arranged alternately along the stem. The most distinctive feature is the intricate network of small veins that lie on either side of the yellowish midrib. These veins appear to be recessed into the surface of the leaf. If the tree is in flower, it can always be identified by the curious white flowers which seem to have 8 petals (more or less) arranged in a rectangular array and with a prominent green ovary in the middle. The pale green center is not round as in most symmetrical flowers, but shaped like a small rounded bar, about twice as long in one diameter as in the other. The flowers are about ½ to ¾ inch across. The fruit is a small black berry less than ½ inch in diameter.

JOHN HOOVER

Look for: A tree of moist to wet forests with dark green, alternate leaves bearing an intricate network of small veins. If present, the white, 8-part flowers are diagnostic.

The wood of kāwaʻu was used for saddle trees (in post-contact times), canoe trim, and for the anvils on which kapa cloth was beaten.

KŌLEA

Myrsine spp.
Myrsinaceae (ardisia, Hilo holly)

There are 20 endemic species of kōlea in Hawaiʻi, over half of which occur only on Kauaʻi. They can be divided into the large-leaved species, which are shrubs or small trees, and the two small-leaved species which are shrubs. I will describe *M. lessertiana*, kōlea lau nui (large leaf kōlea), since it is the most widespread of the tree kōlea, and once you are familiar with this, you should be able to recognize the other tree species as well. Then I will mention the kōlea lau liʻi, or small leaf kōlea, which are shrubs.

Look for: A shrub or tree with colorful young leaves unfurling from a sharp conical bud and with small woody knobs along the sides of older twigs.

Kōlea lau nui is a tree of intermediate to wet forests. It has alternate oval leaves that are 2 to 5 inches long, bluntly pointed, and often broader near the tip than near the base. Young leaves emerge from the tip of the twig in a pointed cone, which is often quite long and slender. As they unfurl from this cone, the leaves are commonly brightly colored, with shades ranging from an unattractive dirty greenish-pink to bright pink, lavender, and maroon. A cluster of these young leaves can be dazzling, rivaling the showiest blooms of the forest in beauty. The stem and midrib of the leaf often retains a deep reddish color in maturity. The small flowers cluster along the sides of the twig on small woody knobs that remain after the fruit has dropped off. The fruit are black berries about ¼ inch in diameter.

Look for: A shrub with inch-long leaves with woody knobs at the base of each, in mesic to wet forests.

The wood of the kōlea was used for house construction by the Hawaiians, and made into anvils or beaters for beating kapa. The red sap of the tree was used as a dye for coloring kapa, and its charcoal was made into a black dye.

Kōlea lau liʻi, is a shrub with spoon-shaped leaves an inch or less long. One species, *M. punctata*, is found on Kauaʻi and very rarely on Oʻahu. It is distinguished from the species found on the other major islands in having a few small teeth near the tip of the leaf. *M. sandwicensis* is the more wide spread kōlea lau liʻi. The leaf stems of this plant may be purple, a color that extends in a sharp triangle into the base of the leaf. Leaves are dark green above and pale gray-green beneath, with a darker midrib running from the stem to about ¾ of the length of the leaf. Note the woody knobs at the base of each leaf. These persist on the stem after the leaf has dropped. This kōlea appears to hybridize with its larger relatives, so that intermediate forms may be seen. The plants are found in mesic to wet forests on all the larger Islands.

LOULU
Pritchardia spp.
Arecaceae (coconut and other palms)

There are about 19 endemic species of this palm found in Hawaiʻi, each restricted to one island, with another 6 located on islands in the southwestern Pacific. Pollen deposits suggest that these palms were once common in coastal regions and extended from there to high in the wet forest. Now, on

O'ahu at least, you are most likely to see them near the wet summits of our higher mountains. These palms are medium sized fan palms, 20 to 30 feet tall, with smooth gray or tan trunks showing rings where fronds have fallen off, leaving a slight leaf scar. The leaves are palmately compound palm fronds, that are very large and stiff with strong spineless stems.

JOHN HOOVER

Look for: A middling sized fan palm with a slightly ringed, smooth trunk, and a tuft of fan-shaped fronds on top, especially seen on wet, wind-swept mountain slopes.

The Hawaiians do not seem to have made much use of these palms. The leaflets may occasionally have been plaited to produce mats. The maka'āinana, or commoners, had a temple, the heiau māpele, in which Lono was worshiped and prayers and offerings were made to insure adequate rainfall and good crops. In season, a small heiau loulu was constructed of loulu fronds in this temple and the gods that presided over fishing were propitiated in it.

Rats gnaw open the nuts of the loulu and destroy most of the seeds before they can germinate. In many areas there are few young trees or seedlings, and the future of these plants in the wild is very uncertain.

MAI'A, BANANA
Musa x paradisiaca
Musaceae (bananas)

This plant was introduced by early Polynesian settlers in the Islands, and although it was never a major food crop here, many varieties were developed. The plant is native to India, south-east Asia, northern Australia, and adjacent islands. It is a hybrid between *M. acuminata* and *M. balbisiana* and is generally sterile, being propagated from the shoots that appear at the base of the plant. While it can grow to be as tall as a small tree, mai'a is actually a giant herb—the trunk is not woody but composed of overlapping layers of spirally arranged leaf bases. It is about 80% water. Mai'a are often found in wet, sheltered valleys at lower elevations. The Hawaiians apparently planted them widely in such situations as an emergency food reserve, but they were also found along the walls between taro paddies and as windbreaks next to houses. Most of the varieties were best when cooked, though some could be eaten cooked or raw and a few were generally eaten raw. Few people can be

191

unfamiliar with the banana tree. It is shaped much like a small coconut palm but lacks the woody trunk. The leaves are very large, as much as 8 to 12 feet in length and 2 feet broad, although those in the wild are generally much smaller. If there is any wind at all, the leaves are likely to be split along the side veins, so that in windy areas, the leaf soon comes to look like a feather. A single flowering stalk emerges from the center of the crown and usually bends over to bear the bunch of bananas arranged in hands. A long tail with a large, reddish-black bud at its end usually dangles below the last hand. A few varieties bear fruit on erect stalks, and one, the maiʻa hāpai (pregnant banana) develops its fruit inside the trunk, which splits open to reveal a few small, sweet bananas inside.

Look for: A banana tree—a large, palm-shaped, soft-trunked herb which may bear a bunch of bananas.

Kāne and Kanaloa, 2 of the 4 major gods of the Hawaiian pantheon, were believed to have planted maiʻa, and the plant was thought to be one form (kino lau) of Kanaloa. Only 3 of the many dozens of varieties of maiʻa were foods that women were allowed to eat. All the rest were kapu. Angela K. Kepler, in *Hawaiian Heritage Plants*, tells the story of how the young Princess Kapiʻolani and her friend Keoha violated the kapu by eating a forbidden variety of banana. They were seen, but being royalty, could not be put to death as lesser women would have been. Someone had to be punished, however, so their tutor was drowned instead. This incident may have influenced Kapiʻolani in later years when she became a high chiefess and a Christian and helped to abolish the old kapu system and belief in the Hawaiian gods by defying Pele on the rim of Halemaʻumaʻu and eating ʻōhelo berries without tossing a handfull into the fire pit first.

A number of sayings referred to maiʻa—to say that a man was like a banana trunk was to imply that he looked big and sturdy, but could be blown over by a puff of wind.

Maiʻa had many uses for the Hawaiians. The trunks and broad leaves were used to line the imu and keep dirt out of the food, as well as providing water to steam it. Maiʻa bunches could be offered on the altars of certain gods at particular times in lieu of human sacrifices. Nectar from maiʻa flowers was given to infants to drink. Different parts of the plant were used in a number of medicinal recipes to treat such maladies as thrush, heartburn, tuberculosis, constipation, asthma, listlessness, and chest pains. Maiʻa fiber was used in weaving and hula skirts, and the dried leaf bases were used to line the thatch on the inside of the roof of a house.

MICONIA
Miconia calvescens
Melastomataceae (clidemia, glorybush, oxyspora)

Miconia is probably the most menacing of all the potentially invasive weeds that threaten the native ecosystems of the Hawaiian Islands. The tree is native to Central and South America and was brought into Hawaiʻi by the nursery trade as an ornamental in the 1960s. It was first reported in the wild in 1982. The plant has devastated vast areas of forest in Tahiti and is a major problem in Sri Lanka, where it aggressively invades native forests or disturbed areas and forms dense monocultures in which nothing else will grow and which provide no suitable habitats for native animals or birds. The trees have shallow roots, and on steep slopes are unable to hold the soil during the heavy rains typical of the wet season in the tropics. As a result, the tree promotes landslides and erosion. Its presence on Kauaʻi and Oʻahu is limited. Mature plants were established in nurseries in Honolulu and the seeds have been spread, so that young plants are often found, sometimes in very remote areas. The seeds will persist in the soil for ten years or so, and it will be necessary to search for this plant and eradicate it whenever it is found for many decades. We can never be sure that mature, fruiting trees have not become established in inaccessible hanging valleys on steep cliff sides or in other remote locations that will continue to serve as a source of seeds. Unfortunately, extensive groves of miconia have flourished on Maui and Hawaiʻi for many years, and the problem of eliminating this pest from these islands is even more severe.

Thomas H. Rau

Look for: A tree with very large, melastome-type leaves, with 3 prominent veins running longitudinally from stem to leaf tip, that are usually reddish-purple underneath.

Miconia is one of the easiest plants to recognize. It is found in moist to wet forest. Trees can grow up to 50 feet in height. The leaves are opposite and

193

very large—up to 2.5 feet long, and, as typical melastomes, have 3 prominent veins running the length of the leaf from base to tip. The upper surface of the leaf is dark green, but the lower surface is often a reddish-purple in color. Innumerable small round purple fruit are widely dispersed by birds and rats.

If you should find a miconia tree on Oʻahu or Kauaʻi, or an isolated plant on one of the other islands, pull it up if possible. Whether you can destroy the tree or not, make a note of its location (by GPS if possible) and report this to the Hawaiʻi Department of Agriculture, Plant Pest Control branch, (808) 973-9541.

ʻOHE, ʻOHEʻOHE

Tetraplasandra oahuensis
Araliaceae (octopus tree, panax, English ivy, ginseng)

KEN SUZUKI

There are 6 endemic species of *Tetraplasandra* in Hawaiʻi, and one endemic species of a close relative, *Reynoldsia*. Both are often called ʻohe, a name also used for bamboo, which they do not resemble at all. ʻOhe are generally 20 to 30 feet tall or less, with alternate odd-pinnate compound leaves. Each leaflet is 4 to 6 inches long by 1 to 3 inches wide and is thick, leathery, and stiff.

Look for: Small trees of mesic to wet forest with odd-pinnately compound leaves, with 5 to 7 large leaflets.

There are usually 2 or 3 pairs of leaflets plus the terminal one. The leaves tend to be clustered at the end of the twig. Flowers form erect umbrella-shaped clusters at the end of the branch and are small and yellow green or pinkish in color. The ripe fruit is deep purple, about ¼ inch in diameter, and may appear to have a recessed cap at the top. These trees are found in mesic to wet forest on all the major Islands.

Reynoldsia is similar except that it is found in dry to mesic forest on all the larger Islands except Kauaʻi, and has leaflets with prominent, usually rounded, teeth along each edge. This tree sheds its leaves during the dry season.

I have been told that Dr. Wayne Gagne, who studied the insects associated with these trees, believed that they were among the earliest colonizers of the Islands because of the large number of native insects that they were host to. A rare related endemic species, *Munroidendron racemosum*, may be seen at Kōkeʻe on Kauaʻi and has been planted occasionally as an interesting garden tree. It has a long, dangling flower spike, up to 3 feet long, with whitish flowers and purple fruit.

'ŌLAPA, LAPALAPA
Cheirodendron spp.
Araliaceae (panax, English ivy, ginseng, octopus tree)

This is a genus of trees found in Polynesia, with 5 endemic species native to Hawai'i, and one that is endemic to the Marquesas. They are small trees to about 30 feet in height with opposite, palmately compound leaves with 3, or sometimes 5, large oval leaflets, 2 to 6 inches long, branching from a common point on the main leaf stem. Clusters of small green and purple flowers are followed by dark purple fruit that is round or 3-sided and about ¼ inch in diameter. Three of the 5 species are found only in mesic to wet forest on Kaua'i, but the most widely spread species, *C. trigynum*, occurs in mesic and wet forest on all the higher Islands. All of these are called 'ōlapa.

C. platyphyllum is found only on Kaua'i and O'ahu in very wet forest. This species is called lapalapa. It has leaflets that are broader than they are long, with each leaflet reminding me of an aspen leaf in form and in the way it quivers in the wind.

Look for: Fairly common small trees of the upper mesic to wet forest with palmately compound leaves with 3 to 5 large oval leaflets.

A bluish dye for staining kapa cloth was obtained from the fruit, leaves, and bark of these trees, and the leaves were used in lei. The tree gives off a strong carrot-like odor when cut, and the bark was used to scent māmaki kapa. The wood is reputed to burn when green, and was therefore especially valued by parties forced to spend the night in the cold, wet mountains, such as the bird catchers who trapped the small native birds for the feathers used in the cloaks of the ali'i. Hula dancers were encouraged to let their hands emulate the grace of the leaves of the 'ōlapa as they trembled in the breeze.

VINES

HOI, PI'A
Dioscorea bulbifera, D. pentaphylla
Dioscoreaceae (yams)

The Polynesian settlers of Hawai'i brought 3 species of yam to the Islands. One of these, *D. alata*, was grown extensively as a food crop in certain areas, particularly on Ni'ihau. In the early days of Western contact, sea captains purchased many tons of these yams to feed their crews. Yams were preferred to sweet potatoes because they had less tendency to sprout while in storage aboard the vessels. However, this domesticated yam does not survive on its own in the wild, and so is not seen along our trails.

Hoi, *D. bulbifera*, is a vine found in shady moist valley forests. It coils to the left to twine around the plants that support it. Its leaves are heart-shaped, 2 to 8 inches long, and with 7 to 11 prominent veins running from base to tip. Small round woody bulbs are often found on this vine, attached to the stem where the leaf stalk joins the vine. There are no prickles on hoi.

Look for: A vine with a heart-shaped leaf with prominent veins and small woody tubers.

Pi'a, *D. pentaphylla*, is a much less common vine that is found in similar habitats. It also coils to the left to twine around the plants on which it clambers. It has a palmately compound leaf with 3 to 5 leaflets radiating from a single point at the end of the stalk. Each broadly spindle-shaped leaflet is 2 to 4 inches long and may be covered with yellow-brown hairs, especially on the lower surface. Small round to horse shoe shaped tubers may

Look for: A vine with a 5 fingered leaf and small woody tubers attached near the base of the leaf stalk.

be attached to the vine at the point where the leaf joins it. The stem usually has scattered prickles on it.

Both hoi and piʻa are native to tropical Asia and were introduced to the Islands by the Hawaiians, probably as an emergency source of food. Neither of these yams were cultivated, but were apparently planted in the mountains and allowed to grow wild. Both plants produce starchy underground tubers in addition to the aerial tubers seen along the vine. It is the underground tuber that was consumed. Piʻa is edible, but apparently was only eaten when nothing better was available. Hoi is poisonous and acrid in taste and must be boiled in lime with frequent changes of water to make it edible.

FERNS

GONOCORMUS
Gonocormus minutus
Hymenophyllaceae (vandenboschia)

I include this fern because it is one of those tiny gems of the forest that is worth notice, but is often overlooked. It is a delicate fern, usually found growing among mosses on boulders in moist shady areas. The fronds are fan-shaped, often forming a nearly complete circle, and depressed in the center to make a shallow cone. They have lobes along the edges. A typical gonocormus is about ⅜ of an inch in diameter. I have been told that this is the world's smallest terrestrial fern, although none of my sources confirm this. They do state that it is the small-

Look for: A tiny, shallowly cone-shaped fern with lobed edges on moss-covered boulders in moist, shady gulches. Note the trumpet-shaped fruiting bodies. Look closely! Think small!

est fern in Hawai'i however! One or two tiny trumpet-like structures are often seen near one edge of the frond. These are the spore-bearing bodies. This fern is thought to be indigenous, although its classification is somewhat controversial, and it also occurs in east Asia and adjacent islands. It is found in Hawai'i on all the major Islands in lowland moist shady gulches.

MAIDENHAIR FERN, 'IWA'IWA
Adiantum raddianum
Pteridaceae (gold fern, Cretan brake, doryopteris)

There are 5 species in this genus in Hawai'i, one rare indigenous species and 4 introduced. The one above is an alien species that has become widespread and is the one that you are most likely to see. The fronds of this fern are 10 to 16 inches long and the overall form is roughly triangular. The frond is highly branched with the terminal leaflets being fan shaped, with rounded lobes fringing the broad terminal ends of the fan. The sori, or spore-bearing bodies, are found on the tips of these lobes and are small dark and kidney-shaped. This is an important feature in distinguishing this fern from the less common native, *A. capillus-veneris*, which has sori that are rectangular bars. This maidenhair fern is native to the American tropics and was first collected in Hawai'i in 1910. It is common on moist banks along our trails on all the major Islands.

The indigenous *A. capillus-veneris* is similar enough to this introduced species that the newcomer was given the same name by the Hawaiians, 'iwa'iwa, which comes from the name of the frigate bird ('iwa). An ancient chant likens the grace of the hula dancers to that of the soaring frigate bird or the waving fronds of a bank of ferns in the breeze. The native fern seems to be increasingly rare in Hawai'i, although it is also found throughout the world in the tropics and subtropics. The

BRADLEY F WATERS

Look for: A typical maidenhair fern with much divided fronds that end in fan-shaped leaflets with rounded terminal lobes and kidney-shaped spore-bearing sori.

Hawaiians made limited use of the fine black shiny but brittle stems as accents in the design of lau hala purses and other artifacts.

MULE'S FOOT FERN
Angiopteris evecta
Marattiaceae (no familiar relatives)

This is a huge terrestrial fern which may be 2 feet thick at the base and sends up enormous spreading fronds that tower over surrounding plants. The frond

stalks may be 2 inches or more in diameter and over 20 feet long. The fronds, which may be 9 feet wide, are twice divided and there is a prominent swelling of the secondary stem where it meets the main stalk. The spores are borne in special structures that form lines near the edges of a leaflet. The leaflets may be 10 inches long and taper to a sharp point. The plant is native to Madagascar, tropical Asia, and the western Pacific. It was introduced into Hawai'i by Lyon Arboretum in 1927 and is now spreading out of control in moist

shaded valleys and adjacent ridge sides where it is replacing native vegetation on all of the major Islands except possibly Kaua'i. It is easily seen along the Mānoa Falls trail on O'ahu.

The crushed leaves of this fern are said to have a maile-like fragrance. There is a native fern, *Marattia douglasii*, that is in the same family as the mule's foot fern (and is also called "mule's foot fern") but it is much smaller and fairly rare, so that you are unlikely to encounter it.

Look for: Once seen, you cannot mistake this fern. No other fern approaches it for sheer size. Although it has no trunk, the massive base and long, thick frond stalks make it far larger than any of our tree ferns.

SELAGINELLA, LEPELEPE A MOA, SPIKE MOSS
Selaginella arbuscula
Selaginellaceae (*Selaginella* is the only genus in this family)

The selaginellas are small, rather fragile, lacy fern-like plants that grow where it is shady and damp. They have many branches which generally lie in one plane so that the plant may resemble a fern frond. The leaves are scale-like, round to linear, with a single vein, and usually only a fraction of an inch long. In some species, the leaves are all alike and spiral around the stem, while in others two kinds of leaves are arranged in alternate rows, the larger, lateral ones lying opposite each other in a plane while the smaller ones hug the stem. These are rather primitive vascular plants that are classified as "fern allies".

The species above is common in Hawai'i and may be endemic, but there is a possibility that it occurs in the Marquesas Islands also and would then be considered indigenous. This plant has 2 rows of lateral leaves on opposite sides of the stem with rows of smaller leaves lying between these and clinging to the stem. The tips of the stem bear clusters of small pointed leaves that bear the reproductive, spore-producing, structures of the plant. This selaginella is common on shaded soil, rocks, or cliffs in damp areas of mesic and wet forest.

Look for: Small delicate fern-like plants with many branches that usually lie in a plane and tiny leaves of two kinds in alternate rows along the stem. The Hawaiian name, lepelepe a moa, means the "comb of a chicken".

VANDENBOSCHIA
Vandenboschia spp.
Hymenophyllaceae (gonocormus)

These ferns belong to a family often called the "filmy ferns" because they are so thin, fragile, and dependent on very moist conditions. There are 4 endemic species in this genus in Hawai'i. They are all ferns of dark wet gulches, heavily shaded rainforest, and shaded dripping cliff faces. The lacy fronds are triangular to lance shaped, divided 2 or 3 times, fragile, translucent, and green to gray-green in color. Between the veins, the fern blade is only one cell layer thick—a thin film indeed! A typical frond is 4 to 18 inches long by 2 to 7 inches wide. Small irregular teeth may be found along the edges of a leaflet and the trumpet-shaped spore-bearing structures occur at the tips of the lobes. Vandenboschia can be found in appropriate habitat on all the major Islands.

Look for: Fragile-looking lacy ferns with translucent fronds in dark moist habitats.

WAHINE NOHO MAUNA
Adenophorus tamariscinus
Grammitidaceae (no familiar relatives)

There are 9 endemic species in this genus in Hawai'i plus a number of hybrids between two species. The Hawaiian name for this fern means "woman who lives on the mountain", and a well dressed lady she is too, with beautiful, delicate, intricately patterned fronds that are always a delight to behold when you find her sitting on a mossy branch high in the wet forest. This is a small fern, about 3 to 8 inches tall, with twice divided, pinnately compound fronds in which the secondary stems branch alternately from

Look for: A small fern perched among mosses on tree limbs in the wet forest. The frond is dagger-shaped, divided into a very regular lacy pattern of tiny leaflets.

Thomas H. Rau

201

the main stalk and each of these bear a row of tiny oval leaflets on each side. The overall shape of the frond is like a dagger blade. Spore-bearing sori occur near the tips of the tiny leaflets as small round dots that are often wider than the leaflet itself. This common fern grows in clusters on tree branches or on the ground in shady wet forests at about 1000 to 4500 feet in elevation.

WĀWAIʻIOLE, LYCOPODIELLA, CLUB MOSS
Lycopodiella cernua
Lycopodiaceae (lycopodium)

This plant is an indigenous fern-ally that is found throughout the tropics and subtropics world wide. It has upright stalks to about 3 feet high with many branches thickly clothed with small needle-like yellow-green leaves that spiral around the stems. The plant often resembles a miniature Christmas tree, and it is sometimes made into wreaths at Christmas time. It is also occasionally used in lei. The leaves point upward or outward, and are about ¼ inch long, with the tips curved inward. The reproductive structures consist of scale-like leaves which cluster at the branch tips and often droop. The Hawaiians thought that these resembled the feet of a rat, and wāwaiʻiole means "rats foot". The plant is common in moist mesic and wet forest and bogs on all of the major Islands.

This plant was boiled for 3 hours by the Hawaiians to produce a solution that was used to bathe the painful joints of people with rheumatism. The abundant giant ancestors of these plants, hundreds of millions of years ago, dominated the ancient forests and enormous quantities of the fallen trees piled up into thick layers that were buried and compressed until they turned into the immense coal seams that we mine today.

KEN SUZUKI

Look for: Plants like miniature Christmas trees about 3 feet tall with branches thickly clothed in needle-like yellow green leaves.

Subalpine Zone

HERBS

CATCHFLY, CAMPION
Silene struthioloides
Caryophyllaceae (carnation, pink, sweet william, baby's breath)

This is one of 7 species in this genus that are endemic to Hawai'i, and is probably the one you are most likely to see as it is often found along the trail in Haleakalā on Maui, while most of the other species are rare or extinct. The plant is a many-branched, diminutive shrub. It has small, awl-shaped leaves about ⅓ inch long that lie in tight clusters along the stem. The flowers, if present, are white, about ⅝ inch in diameter, and have 5 narrow petals well separated from each other but broadening at the tip which is divided into 2 equal lobes. The plants are found scattered in alpine and subalpine shrublands in Haleakalā and on Mauna Kea on Hawai'i.

Look for: Small branching plants with small awl-shaped, stem-hugging leaves and 5- petalled white flowers with each petal broadening at the tip into 2 lobes.

A close relative, *S. hawaiiensis*, is widely distributed on the Big Island and may be seen around Kīlauea caldera in Hawai'i Volcanoes National Park.

EVENING PRIMROSE
Oenothera stricta
Onagraceae (fuchsia)

These plants are slender erect hairy herbs 1 or 2 feet tall, with narrow, stemless, alter-

Look for: Slender erect herbs with showy 4-petalled yellow flowers that open in the afternoon along our National Park trails.

203

nate leaves 1 to 4 inches long. They have shallow teeth along each edge. The 4-petalled yellow flowers, about 2 inches in diameter, open in the afternoon and turn reddish and wilt by the next morning. This ornamental is native to Chile and Argentina and was first collected on Maui about 1919. It was probably introduced as a garden flower. It is especially common along trails in Haleakalā and Volcanoes National Parks.

JOHN HOOVER

SILVERSWORD, ʻĀHINAHINA
Argyroxiphium spp.
Asteraceae (lettuce, ragweed, dahlia, sunflower)

There are 5 species in this endemic genus, 3 silverswords and 2 greenswords. The one that you are most likely to see is *A. sandwicense*, also called ʻāhinahina or just hinahina. Since these are names the Hawaiians applied to a wide variety of gray and silver colored plants, they can be confusing and we will use the post-contact common name, silversword. The *Manual* calls this plant a shrub, but it is a ball-shaped plant that hugs the ground and is never much more than knee high until it flowers, when a stalk that may become 6 or 7 feet tall is produced. The leaves of this plant are succulent, containing a gel that helps it store water, a useful trait in its open arid windswept alpine environment. They form a globular basal rosette that may reach nearly 2 feet in diameter. Each leaf is a curved rod, triangular in cross section, that tapers to a blunt point, and is 8 to 16 inches long by half an inch wide. They are covered with a dense mat of fine, silky white hairs that provide the plant with its distinctive silver color. The cluster of leaves surround the stem to form ball with a depression on top where the young leaves are emerging. A view from above into the heart of this globe, when the silvery-green young leaves are glowing in the sun in a regular geometric array, is one of the loveliest sights in the plant kingdom. This plant generally blooms only after years of slow growth—perhaps up to 20 years in some cases. It blooms only once and then dies. The flowering stalk on a large, healthy silversword will bear hun-

Look for: A lovely, shining silver ball composed of curved, sword-blade leaves rising from its base and enclosing a beautiful silver-green, geometrically patterned array of young leaves.

dreds of sturdy sunflower-like heads, up to an inch and a half in diameter, with rather scanty, widely spaced pink to wine-red petals. Flowering occurs from mid-June to November. The plant can generally be seen at the Visitors Center of Haleakalā National Park on Maui, where a few are usually kept planted by the front door. Hikers will find it growing in abundance on alpine cinder along the mid section of the Sliding Sands Trail into the Crater, and on the Silversword Loop on the trail from Hōlua Cabin to the Crater center. A small population also occurs on Mauna Kea on the Big Island (Hawai'i).

The other 2 species of silverswords are similar in appearance. The 'Eke silversword, *A. caliginis*, occurs in bogs on the high peaks of West Maui and the Ka'ū silversword, *A. kauense*, is also a bog and wet forest plant found in very limited areas on the southeast slope of Mauna Loa. The greenswords are rare—*A. virescens* is probably extinct—and these plants are more shrublike. *A. grayanum* is an erect shrub with green, narrow, sword-like rosettes of leaves. It is found in wet, high forests and the edges of bogs on Maui.

Argyros means "silver" in Greek, and xiphium is dagger, so the scientific name is much the same as the common one. These plants were pushed to the brink of extinction by the browsing of goats and pigs, but are recovering nicely where these animals have been fenced out of the National Parks and other protected areas. They are also threatened by the alien Argentine ant, however, which is beginning to invade Haleakalā and which preys on other insects, including the native moth that pollinates the silversword.

TETRAMOLOPIUM

Tetramolopium humile
Asteraceae (marigold, thistle, lettuce)

One rare (perhaps extinct) Hawaiian species in this genus was called "pāmakani", but since this name was also used for several other plants, and was not applied to this species in any case, it seems best to use the scientific name in the absence of a better alternative. There are 36 species in this genus and they are found only in the Cook Islands, New Guinea, and Hawai'i. The 11 Hawaiian species are all endemic, except for *T. sylvae*, which is indigenous, being

also found in the Cook Islands. All the Hawaiian species are plants of dry areas found in a range of elevations extending from sea level to about 10,000 feet. All but the one above are rare or extinct. *T. humile* is a small woody herb that branches from the base. The small, alternate, narrow leaves tend to lie parallel to the stem and form a sheath around it. The flowers are about ½ inch across, white to lavender, and daisy-like. The plant is found in alpine deserts in crevices in lava flows or on cinder slopes. It is not uncommon on Haleakalā on Maui and on the higher mountains on the Island of Hawai'i.

Look for: A small herb of open alpine scrubland with small linear leaves sheathing the stem and daisy-like flowers.

SHRUBS

GERANIUM, HINAHINA
Geranium spp.
Geraniaceae (cranesbill, pelargonium)

Six endemic and 5 naturalized species in this genus have been reported from Hawai'i. All but one of the natives are restricted to a single island and are rare and unlikely to be encountered. The one that you are most likely to see, hinahina or *Geranium cuneatum*, is found in high altitude scrubland on Haleakalā, Maui, and on the Big Island. Since "hinahina", meaning "gray haired" or "silvery" , is used for several other completely different plants, I will call the plants "geranium". This geranium is an attractive, erect, silvery-leaved compact shrub with many branches. The

Look for: An attractive, compact shrub of high altitude scrubland on Maui and the Big Island with silvery inch-long foot-shaped leaves and 5-petalled white flowers.

leaves are alternate, about 1 inch long, and roughly shaped like a foot, being broader near the tip than at the base, with 3 to 8 triangular teeth representing the toes. Prominent parallel veins run from base to tip of each leaf. Soft white hairs cover the underside of each, giving the leaf its silvery hue. The showy 5-petalled flowers are about ¾ inch in diameter, white, often with pink or purple veins or throat.

NA'ENA'E, KŪPAOA
Dubautia spp.
Asteraceae (cocklebur, zinnia, chrysanthemum, daisy)

This whole genus, containing 21 species is endemic to Hawai'i, with almost half of the species being found only on Kaua'i. These are highly variable plants, ranging from herbs to shrubs to small trees and vines. I will discuss 3 fairly wide spread or noticeable species that illustrate the great differences in form exhibited by these plants. Once you can recognize these, you should be able to tell that the remaining species are near cousins to one of them. All of the Asteraceae have composite flowers—that is, what appears to be a single flower is actually a collection of a few to many flowers enclosed in a single floral cup. In the naʻenaʻe the flower heads are usually small and numerous, and you will have to look closely to see the individual flowers within each flower head.

D. plantaginea is a shrub or small tree. Its opposite leaves are long, pointed ovals, 3 to 10 inches long by ½ to 2 inches wide and have 7 to 16 parallel veins running the length of the leaf. There are small teeth on the edge of the leaf from midpoint to the tip. There is no obvious leaf stalk, but the base of the leaf seems to merge with the branch or even to clasp it. Rings are left behind on the plant stem where old leaves have fallen off. The flowers are very numerous and form fuzzy clusters on stalks at the tips of the branches. They are a drab yellowish to purple in color with no petals. This plant is found in mesic to wet forest on all of the larger Islands.

Look for: A shrub with opposite leaves with parallel veins and no leaf stalk. Flowers are clusters of small, fuzzy, dull colored, petalless blossoms.

D. menziesii is common along the Sliding Sands trail in Haleakalā, a walk that every hiker in the Islands should take for its spectacular other-worldly scenery. The plant is a small to fairly large shrub up to 7 feet tall, with rigid, woody, erect stems. The leaves are about 2 inches long with the shape of somewhat rounded elongated triangles, stiff and thick, and are often arranged in regular rows along the stem. If you look down the length of a branch from the tip, you will see that one leaf falls directly behind the one in front in 4, or sometimes 6, rows; the ends of the leaves defining the corners of a square or a regular hexagon. The flowers are open clusters of small fuzzy drab yellow-orange blossoms. This hardy

Look for: A common shrub on Haleakalā with short stiff leaves in uniform rows with 4 or 6 rows along a branch.

plant occurs in subalpine shrubland and alpine deserts on East Maui.

This na'ena'e looks nothing like the Haleakalā silversword and it is hard to believe that they are closely related. Yet they are in fact sufficiently alike genetically that they can cross and form viable hybrids. When I was in Haleakalā in 2004, there were 2 or 3 plants that were clearly hybrids between these 2 species scattered among the silverswords half way down the Sliding Sands trail.

D. scabra is a low growing, spreading, half-herb, half-shrub that often forms a mat. The shoots of the plant are slender and purple in color. The small narrow rough alternate leaves are ½ to 3 inches long with a midrib and tiny teeth. They are often densely clustered but become very sparse near the flowering stalk. The small white blossoms form open clusters on long, slender stems. This na'ena'e is found on recent lava flows and in wet forest on Moloka'i, Lāna'i, Maui, and Hawai'i.

Look for: A low growing, spreading plant with small, narrow, alternate leaves and open clusters of small composite-type white flowers.

The roots and leaves of certain na'ena'e were used to scent kapa by the native Hawaiians.

'ŌHELO
Vaccinium spp.
Ericaceae (heather, azalea, rhododendrum, blueberry, cranberry)

There are 3 species in this genus in Hawai'i, all endemic. They are shrubs of various sizes with alternate oval rounded to pointed leaves which usually have fine teeth along the edges. The small flowers, about ⅓ inch long, are bell-shaped or vase-shaped, waxy, and vary in color from red to yellow-green. Fairly tall wet forest shrubs of 'ōhelo occur occasionally on Kaua'i and O'ahu, but the ones you are most likely to see are low growing bushes on cinder or recent lava fields in Haleakalā, Maui, or around the Kīlauea Volcano on the Big Island. These plants bear round red berries about ⅓ inch in diameter that have a crown-like structure at the end, and which are quite edible and often used to make pies or tarts. Do not eat wild berries lacking this crown! They may be poisonous.

These plants bore one of the few edible native fruits that the Hawaiians found here when they arrived in the Islands. On the Big Island, they believed that the berries were sacred to the Volcano goddess, Pele, and before they would eat any in the vicinity of Kīlauea Crater, they would toss a branch bearing the fruit into the fire pit of Halemaʻumaʻu as an offering to the goddess.

KEN SUZUKI

Look for: Shrubs with leathery oblong to nearly circular alternate leaves with fine teeth and small red berries with a crown at the tip.

The young Princess Kapiʻolani played a major role in overturning the kapu system and clearing the way for the Hawaiian people to accept Christianity by defying Pele on the rim of Halemaʻumaʻu and eating the berries without making such an offering. The failure of Pele to retaliate shook the faith of the Hawaiian observers in their ancient pantheon. See the entry on "MAIʻA" for background on Kapiʻolani's resentment of the kapu system.

TREES

MĀMANE
Sophora chrysophylla
Fabaceae (mimosa, indigo, kidney bean)

Māmane is generally a small, endemic tree with odd-pinnate compound leaves about 5 or 6 inches long and often covered with fine hairs. Each leaf has 6 to 10 pairs of leaflets, plus the odd one at the tip, and each is about an inch long and half an inch wide. Clusters of yellow, inch-long, pea-like flowers are found at the branch tips or near the base of a leaf. The bean pods are 4 to 6 inches long with 4 prominent flanges running the length of the pod, and constrictions between the seeds. The seeds are bright orange. You are most likely to see māmane trees in subalpine vegetation on Haleakalā or the Big Island, where they may be dominant components of the forest.

Look for: A small tree in subalpine habitat with odd-pinnate compound leaves, flanged pods, and yellow flowers.

Hawaiians used the hard durable wood of the māmane for house construction and for the digging sticks or ʻōʻō used by farmers. It was also favored for the runners of the hōlua sleds that the aliʻi rode in their daring sledding contests down prepared chutes lined with grass or kō (sugar cane) flowers and perhaps greased with kukui nut oil. The flowers were used in lei, and they were an important source of nectar for iʻiwi, ʻapapane, and other birds in the honeycreeper family. Unfortunately, the foliage of this tree is highly palatable to sheep and goats, and the survival of the tree where these animals run wild was long in doubt. They have now been fenced out of Haleakalā, however, and also from a large area on Mauna Kea where another rare Hawaiian bird, the palila, depends on the seeds and flowers for food and was in danger of extinction, as the wild ungulates prevented regeneration of the tree. In these areas, māmane is once more reseeding itself, and mature trees no longer show a "browse line" where all foliage is missing below the height that a goat standing on its hind legs can reach. Like koa and aʻaliʻi, māmane can recover after a fire by sending up sprouts from the base of the tree.

SOURCE MATERIALS

Abbott, Isabella Aiona. 1992. *Lāʻau Hawaiʻi: Traditional Hawaiian Uses of Plants.* Honolulu: Bishop Museum Press.

Beckwith, Martha. 1940. *Hawaiian Mythology.* Honolulu: University of Hawaiʻi Press.

Bohm, Bruce A. 2004. *Hawaiʻi's Native Plants.* Honolulu: Mutual Publishing.

Carlquist, Sherwin. 1970. *Hawaii: A Natural History.* New York: The American Museum of Natural History.

———. 1965. *Island Life.* New York: The American Museum of Natural History.

Cassel, Katie. 2005. *Na Pua O Kokee: Field Guide to the Native Flowering Plants of Northwestern Kauai.* Rancho Palos Verdes, CA: Quaking Aspen Books.

Conard, Harry S., and Paul L. Redfearn Jr. 1956. *How to Know the Mosses and Liverworts.* 2nd ed. Dubuque, IA: Wm. C. Brown Publishers.

Culliney, John L., and Bruce P. Koebele. 1999. *A Native Hawaiian Garden.* Honolulu: University of Hawaiʻi Press.

Degener, Otto. 1945. *Plants of Hawaii National Parks.* Ann Arbor, MI: Edwards Brothers, Inc.

Flannery, Tim, and Peter Schouten. 2001. *A Gap in Nature.* New York: Atlantic Monthly Press.

Fornander, Abraham. 1996. *Ancient History of the Hawaiian People.* Honolulu: Mutual Publishing.

Grimaldi, David, and Michael S. Engel. 2005. *Evolution of the Insects.* New York: Cambridge University Press.

Gutmanis, June. 1976. *Hawaiian Herbal Medicine.* Ill. Susan G. Monden. Honolulu: Island Heritage Publishing.

Handy, E. S. Craighill, and Elizabeth Green Handy. 1991. *Native Planters in Old Hawaii: Their Life, Lore, and Environment.* Revised ed. With the collaboration of Mary Kawena Pukui. Honolulu: Bishop Museum Press.

Hargreaves, Dorothy and Bob. 1964. *Tropical Trees of Hawaii.* Portland, OR: Hargreaves Industrial.

Hemmes, Don E., and Dennis E. Desjardin. 2002. *Mushrooms of Hawai'i.* Berkeley: Ten Speed Press.

Hicks, Marie L. 1992. *Guide to the Liverworts of North Carolina.* Durham, NC: Duke University Press.

Jamieson, Dean, and Jim Denny. 2001. *Hawai'i's Butterflies and Moths. A Hawai'i Biological Survey Handbook.* Honolulu: Mutual Publishing.

Juvik, Sonia P. and James O. 1998. *Atlas of Hawai'i.* 3rd ed. Honolulu: University of Hawai'i Press.

Kalākaua, His Hawaiian Majesty King David. 1990. *The Legends and Myths of Hawaii.* Honolulu: Mutual Publishing.

Kamakau, Samuel Manaiakalani. 1991. *Tales and Traditions of the People of Old.* Trans. Mary Kawena Pukui. Bernice P. Bishop Museum Special Publication No. 51. Honolulu: Bishop Museum Press.

———. 1976. *Works of the People of Old.* Trans. Mary Kawena Pukui. Bernice P. Bishop Museum Special Publication No. 61. Honolulu: Bishop Museum Press.

Kay, E. Alison, ed. 1972. *A Natural History of the Hawaiian Islands: Selected Readings.* Honolulu: University of Hawai'i Press.

Kepler, Angela Kay. 1998. *Hawaiian Heritage Plants.* Revised ed. Honolulu: University of Hawai'i Press.

Krauss, Beatrice H. 1993. *Plants in Hawaiian Culture.* Ill. Thelma Grieg. Honolulu: University of Hawai'i Press.

———. 2001. *Plants in Hawaiian Medicine.* Ill. Martha Noyes. Honolulu: The Bess Press.

Kuck, Loraine E., and Richard C. Tongg. 1958. *A Guide to Tropical & Semitropical Flora: Hawaiian Flowers & Flowering Trees.* Rutland, VT: Charles E. Tuttle Co.

Lamoureux, Charles H. 1963. *Field guide to the Maunakapu-Palikea trail.* Hawaiian Botanical Society Newsletter 2: 81-83.

———. 1996. *Trailside Plants of Hawaii's National Parks.* Revised ed. Hawai'i

Natural History Association in cooperation with the National Park Service, U. S. Department of the Interior.

Leary, James J. K., Paul W. Singleton, and Dulal Borthakur. 2004. *Canopy nodulation of the endemic tree legume, Acacia koa in the mesic forests of Hawaii.* Ecology 85(11): 3151-3157.

Lilleeng-Rosenberger, Kerin E. 2005. *Growing Hawai'i's Native Plants.* Honolulu: Mutual Publishing.

Little, Elbert L. Jr., and Roger G. Skolmen. 1989. *Common Forest Trees of Hawaii (Native and Introduced).* Agriculture Handbook No. 679. Washington, D.C.: United States Department of Agriculture, Forest Service.

Malo, David. 1997. *Hawaiian Antiquities.* Trans. Nathaniel B. Emerson. Revised ed. Honolulu: Bishop Museum Press.

Merlin, Mark. 1978. *Hawaiian Coastal Plants and Scenic Shorelines.* Honolulu: The Oriental Publishing Co.

———. 1995. *Hawaiian Forest Plants.* Honolulu: Pacific Guide Books.

Neal, Marie C. 1965. *In Gardens of Hawaii.* Revised ed. Bernice P. Bishop Museum Special Publication 50. Honolulu: Bishop Museum Press.

Palmer, Daniel D. 2003. *Hawai'i's Ferns and Fern Allies.* Honolulu: University of Hawai'i Press.

Pope, Willis T. 1968. *Manual of Wayside Plants of Hawaii.* Rutland, VT: Charles E. Tuttle Co.

Pratt, Douglas H. 1999. *A Pocket Guide to Hawai'i's Trees and Shrubs.* Honolulu: Mutual Publishing.

Pukui, Mary Kawena, and Samuel H. Elbert. 1986. *Hawaiian Dictionary.* Honolulu: University of Hawai'i Press.

Rock, J. F. 1974. *The Indigenous Trees of the Hawaiian Islands.* Reprint ed. Rutland, VT: Charles E. Tuttle Co.

Sohmer, S. H., and R. Gustafson. 1987. *Plants and Flowers of Hawaii.* Honolulu: University of Hawai'i Press.

Staples, George W., and Robert H. Cowie. 2001. *Hawai'i's Invasive Species.* A Hawai'i Biological Survey Handbook. Honolulu: Mutual Publishing and Bishop Museum Press.

Staples, George W., and Derral R. Herbst. 2005. *A Tropical Garden Flora.* Contribution No. 2005-003 to the Hawai'i Biological Survey. Honolulu: Bishop Museum Press.

Sterling, Elspeth P., and Catherine C. Summers. 1978. *Sites of Oahu.* Revised ed. Honolulu: Bishop Museum Press.

Stone, Charles P., and Linda W. Pratt. 1994. *Hawai'i's Plants and Animals.* Ill. Joan M. Yoshioka. Honolulu: Hawai'i Natural History Association, National Park Service, and University of Hawai'i Cooperative National Park Resources Studies Unit.

Tudge, Colin. 2006. *The Tree.* New York: Crown Publishers.

Valier, Kathy. 1995. *Ferns of Hawai'i.* Honolulu: University of Hawai'i Press.

Wagner, Warren L., Derral R. Herbst, and S. H. Sohmer. 1999. *Manual of the Flowering Plants of Hawai'i.* Revised ed. In two volumes. Bishop Museum Special Publication 9. Honolulu: University of Hawai'i Press and Bishop Museum Press.

Walther, Michael. 2004. *A Guide to Hawaii's Coastal Plants.* Honolulu: Mutual Publishing.

Westervelt, William D. 1998. *Hawaiian Historical Legends.* Honolulu: Mutual Publishing.

———. 1998. *Hawaiian Legends of Ghosts and Ghost Gods.* Honolulu: Mutual Publishing.

Whistler, W. Arthur. 1995. *Wayside Plants of the Islands: A Guide to the Lowland Flora of the Pacific Islands—Hawai'i, Samoa, Tonga, Tahiti, Fiji, Guam, and Belau.* Honolulu: Isle Botanica.

Yamaguchi, Mas. 1983. *World Vegetables.* Westport, CT: AVI Pub. Co.

Ziegler, Alan C. 2002. *Hawaiian Natural History, Ecology, and Evolution.* Honolulu: University of Hawai'i Press.

INDEX

Plant Names—Hawaiian:

SCIENTIFIC INDEX

ABOUT THE AUTHOR

Born in Denver, Colorado, John B. Hall acquired there his long-enduring love for mountains and trailside plants. After obtaining a PhD in Biochemistry at the University of California, Berkeley, he went on to become an Assistant Professor of Biochemistry at the University of Hawaiʻi at Maʻnoa, and later a Professor of Microbiology. Hall achieved several Fulbright scholarships and taught in a variety of countries including New Zealand, Nepal, Turkey, and Belize.

Hall has explored Hawaiʻi's trails for more than 40 years. He is a life member of both the Hawaiian Botanical Society and the Hawaiʻi Audubon Society; he has led tours into the Honouliuli Reserve as a docent for the Nature Conservatory of Hawaiʻi, and served on the Citizens Advisory Council in the Division of Lands and Natural Resources unit. Hall also served on the Board of Directors for the Hawaiian Trail and Mountain Club and founded a local hiking group known as "Solemates." In addition to his own love for hiking and trailside plants, Hall and his friends have done a great deal to maintain and re-open old paths to make them available to the public.